# Engineering@work

Steve Cushing

## Case studies



A PEARSON COMPANY

**Published by**
Edexcel Limited
One90 High Holborn
London
WC1V 7BH

www.edexcel.org.uk

**Distributed by**
Pearson Education Limited
Edinburgh Gate
Harlow
Essex
CM20 2JE

www.longman.co.uk

**©Edexcel Limited 2007**

**First published 2007**
British Library Cataloguing in Publication Data is available from the British Library on request

10-digit ISBN 1-84690-166-9
13-digit ISBN 978-1-84690-166-9

This material offers high-quality support for the delivery of Edexcel qualifications.

This does not mean that it is essential to achieve any Edexcel qualification, nor does it mean that this is the only suitable material available to support any Edexcel qualification. No Edexcel-published material will be used verbatim in setting any Edexcel assessment, and any resource lists produced by Edexcel shall include this and other appropriate texts.

Edexcel has no responsibility for the persistence or accuracy of URLs for external or third-party internet websites referred to in this book, and does not guarantee that any content on such websites is, or will remain, accurate or appropriate.

**Copyright notice**
All rights reserved. No part of this publication may be reproduced, stored in a retrieval system, or transmitted in any form or by any means, electronic, mechanical, photocopying, recording, or otherwise without either the prior permission of the Publishers or a licence permitting restricted copying in the United Kingdom issued by the Copyright Licensing Agency Ltd, 90 Tottenham Court Road, London W1P 9HE.

Commissioned by Jenni Johns

Design and publishing services by Steve Moulds of DSM Partnership

Project managed by Gwen Burns

Cover images courtesy of Jupiter Images

Picture research by Thelma Gilbert

Figures 2.1/4.6/5.6/7.2/8.4 and drawings of robotic and human arms on pages 119 and 121 by Oxford Designers and Illustrators Ltd

Index by Julie Rimington

Printed and bound by Graficas Estella, Bilboa, Spain

**Acknowledgements**
The author wishes to thank the following for their assistance:

Mark White and Glen Harding of Helipebs Controls Ltd

Les Ratcliffe and Bill McLundie of Jaguar Cars

Nichola King and Sharon Cruz of Wyndeham Press Group Plc

Mike Satur of Mike Satur

Jonathan Foster and Faye Robinson of Troup Bywaters + Anders

John Gathard, Nicky Taylor and Mark Alexander of Ginsters

Stephen Pearson, Jonathan Foreman and Brett Mason of Ceramic Seals.

The author also wishes to thank Adrian Moss for his technical support in filming the case studies.

# Contents

**About this book** 4

CHAPTER 1 — **Fluid power:** Helipebs Controls Ltd 6

CHAPTER 2 — **The motor industry:** Jaguar Cars 28

CHAPTER 3 — **The print industry:** Wyndeham Westway 48

CHAPTER 4 — **Small engineering company:** Mike Satur 72

CHAPTER 5 — **Building services – lighting design:** Troup Bywaters + Anders 92

CHAPTER 6 — **Food manufacturing:** Ginsters 114

CHAPTER 7 — **Precision engineering:** Ceramic Seals 134

CHAPTER 8 — **Building services – mechanical design:** Troup Bywaters + Anders 158

APPENDIX 1 — **Health and safety** 184

APPENDIX 2 — **Standards: ISO** 190

**Index** 194

# About this book

Welcome to *Engineering@Work Case Studies.* This book provides the background information to selected engineering companies, the people who work in them and some of the theoretical knowledge that is required to work in the industry.

When you first look at this book you will see various features, both in the margins and in the text, that guide you on how to build and develop your knowledge and understanding of engineering.

**cutaway** – a diagram or model of an object with part of the outer layer removed to reveal the interior

Definition boxes in the margin allow you to check your understanding of **keywords**.

Follow the web links in the **Nuts & Bolts** feature to get more information about an area of interest.

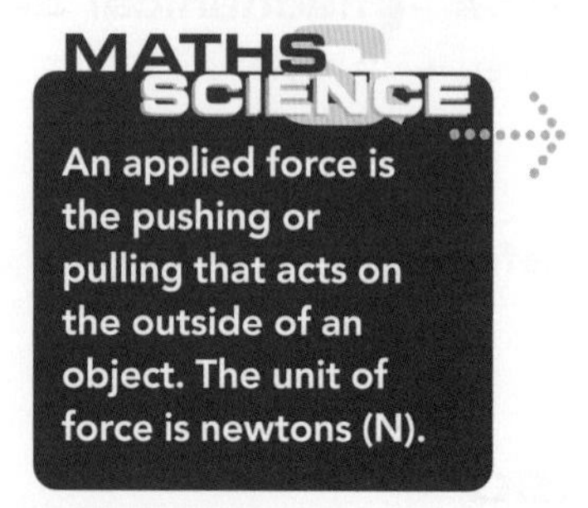


**Maths & Science** boxes in the margin flag up areas in the text where key mathematical or scientific principles are discussed.

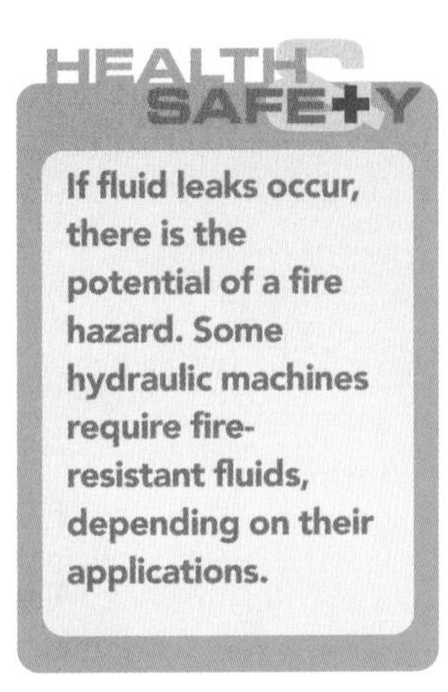


This feature draws your attention to any key **Health and Safety** issues within the text.

The **Career path** provides a profile of the company employee interviewed for the case study and outlines possible courses and career progression that may enable you to begin a career in this field.

Additionally, there are colour photographs and clear illustrations, including technical drawings, to aid your understanding of the underlying engineering theory.

At the end of the book, two appendices are provided with additional information about health and safety, including legislation, within the engineering industry, and the standards that are being set and developed by the International Organization for Standardization (ISO).

We hope you enjoy using this book in your studies, and we wish you the very best both for your course and for your future career in engineering.

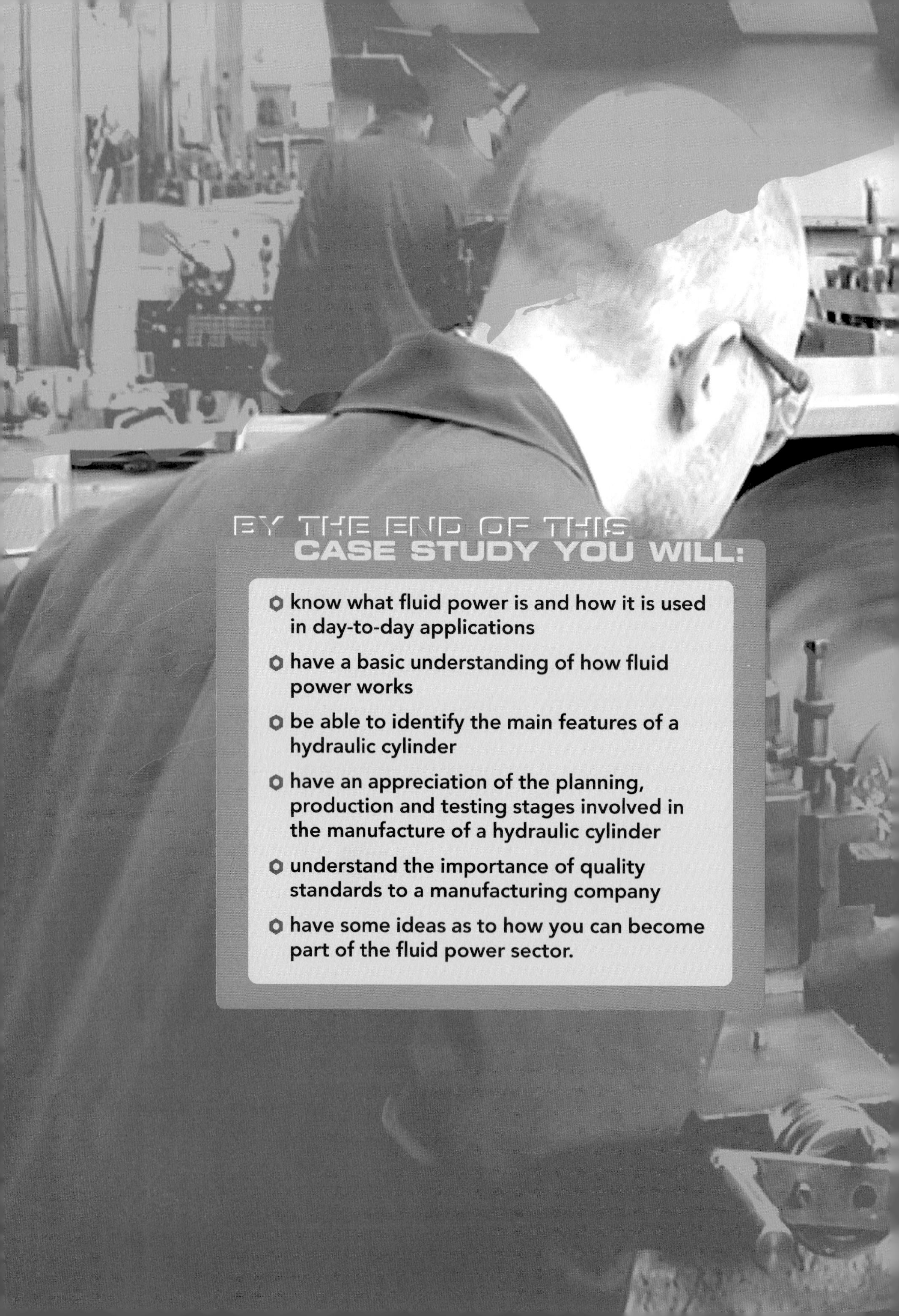

## BY THE END OF THIS CASE STUDY YOU WILL:

- know what fluid power is and how it is used in day-to-day applications
- have a basic understanding of how fluid power works
- be able to identify the main features of a hydraulic cylinder
- have an appreciation of the planning, production and testing stages involved in the manufacture of a hydraulic cylinder
- understand the importance of quality standards to a manufacturing company
- have some ideas as to how you can become part of the fluid power sector.

# FLUID POWER: HELIPEBS CONTROLS LTD

## Case study overview

**This case study introduces you to the world of fluid power and its applications.**

**Over the following pages you will become familiar with a real company that specialises in the production of hydraulic cylinders.**

**After identifying the main elements of a hydraulic cylinder, you will follow the company's processes from initial design idea to the final manufactured product. This will help you to understand all the steps involved and the order in which they must be taken to produce a hydraulic cylinder.**

**Finally, you will be introduced to a design engineer from the company who will describe the route he took to his current position.**

# The sector

**hydraulics** – a system of using liquid to produce power

**hydraulic cylinder** – mechanical actuators that are used to give a linear force through a linear stroke

**piston** – a sliding shaft that fits snugly into a cylinder

## What is fluid power?

Fluid power is energy transmitted and controlled by means of a pressurised fluid, either liquid or gas. The term fluid power applies to both hydraulics and pneumatics. **Hydraulics** uses pressurised liquid, for example, oil or water; pneumatics uses compressed air or other neutral gases.

## Fluid power applications

**Hydraulic cylinders**, as manufactured by Helipebs, are used everywhere. Just look at a few of the applications shown below.

### Aircraft

Flight controls, brakes, thrust reversers, landing gear and wheels require a significant amount of energy, usually hydraulic energy.

### Elevators

Hydraulic elevators use the principle of hydraulics to pressurise an above-ground or in-ground **piston** to raise and lower the car.

### Cars

Brakes and power steering use hydraulics.

### Diggers and farm equipment

Hydraulics is used for lifting, splitting, movement and tipping, and anywhere that a significant amount of energy is required.

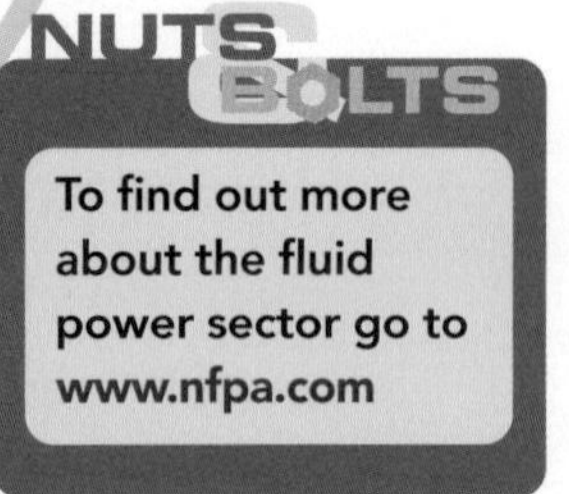


Hydraulic machinery

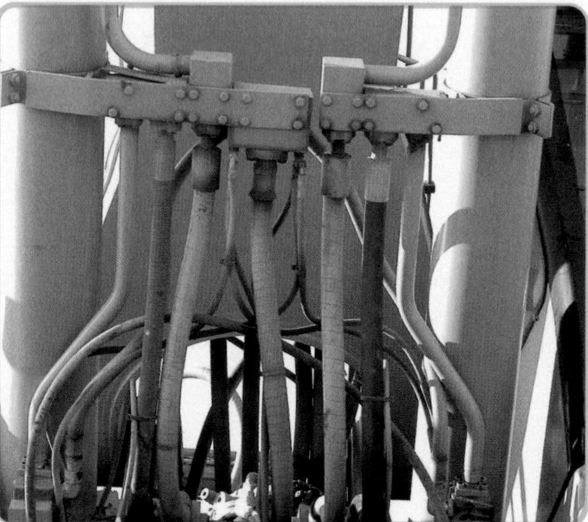

# The company

## Background

Some years before the 1914–1918 war, a cement chemist, William Fennell, began experimenting with HOLPEBS – tubular steel bodies – as a means of improving the fine grinding process in cement mills.

Forged steel balls were used for the initial stages of cement grinding, whilst subsequent processes were still chiefly dependent on the primitive action of flint pebbles gathered from the beaches of France and southern England, or on cylindrical bodies made from mild steel or chilled cast iron.

The first results of Fennell's experiments gave a startling improvement on existing types and, following a further period of research and development, HELIPEBS were introduced. Made from spirally-coiled steel wire, they are tubular like HOLPEBS, and so have a larger grinding surface than the solid cylinder.

Output was increased and steps were taken to make HELIPEBS more widely known. The response was decisive and on 7 July 1922 Helipebs Limited of Premier Works Gloucester was formed.

## Modern day

Today, Helipebs is a medium-sized engineering operation that provides complete hydraulic solutions to end users, who rely on Helipebs' engineering expertise to solve their problems.

**MATHS & SCIENCE**

Imperial units of length are inch, foot and mile. Metric units of length are centimetre, metre and kilometre.

**NUTS & BOLTS**

To find out more about Helipebs Controls Ltd go to www.helipebs.co.uk

Helipebs supply both UK and overseas customers and, because of their American connections, they need to deliver products that conform to both imperial and metric standards.

The industries supplied by Helipebs include:

- petrochemical
- pharmaceutical
- aviation testing.

## Standards

Helipebs has both ISO 9001-2000 and ISO 14001-2004 standards.

**FIG 1.1:** Certificates of registration

Certificate of Registration

QUALITY MANAGEMENT SYSTEM - ISO 9001:2000

*This is to certify that:*

**Helipebs (Holdings) PLC**
**Sisson Road**
**Gloucester**
**GL2 0RE**
**United Kingdom**

*Holds Certificate No:* **FM 14317**

*and operates a Quality Management System which complies with the requirements of BS EN ISO 9001:2000 for the following scope:*

The design and manufacture of a range of hydraulic and pneumatic cylinders, valves, servo-control equipment and accessories. These products may also be manufactured under licence or to customer specifications.
The manufacture of electronic sub-assemblies to customer specification.

*For and on behalf of BSI:*

*Managing Director, BSI Management Systems (UK)*

Originally registered: **17/01/1992** Latest Issue: **10/07/2006**

*Page: 1 of 2*

BSI Management Systems

This certificate was issued electronically and remains the property of BSI and is bound by the conditions of contract.
This certificate does not expire. An electronic certificate can be authenticated online.
Printed copies can be validated at www.bsi-global.com/ClientDirectory

The British Standards Institution is incorporated by Royal Charter.
Management Systems (UK) Headquarters: P.O. Box 9000, Milton Keynes MK14 6WT. Tel: 0845 080 9000

Certificate of Registration

ENVIRONMENTAL MANAGEMENT SYSTEM

*This is to certify that:*

**Helipebs (Holdings) PLC**
**Sisson Road**
**Gloucester**
**GL2 0RE**
**United Kingdom**

*Holds Certificate No:* **EMS 69317**

*and operates an Environmental Management System which complies with the requirements of BS EN ISO 14001:2004 for the following scope:*

The manufacture of hydraulic and pneumatic cylinders, valves, and electrical control systems. The manufacture of grinding media.

*For and on behalf of BSI:*

*Managing Director, BSI Management Systems (UK)*

Originally registered: **05/03/2004** Latest Issue: **13/01/2006**

*Page: 1 of 2*

BSI Management Systems

This certificate was issued electronically and remains the property of BSI and is bound by the conditions of contract.
This certificate does not expire. An electronic certificate can be authenticated online.
Printed copies can be validated at www.bsi-global.com/ClientDirectory

The British Standards Institution is incorporated by Royal Charter.
Management Systems (UK) Headquarters: P.O. Box 9000, Milton Keynes MK14 6WT. Tel: 0845 080 9000

# The product

## Introduction

Fluids can be used in a large number of control systems from robots to cars. A fluid is a continuous shapeless substance. Fluids have a tendency to take on the shape of the container they are held in. Although fluids have no power, they can be confined and pressurised in order to store and transmit power; because of this they are used in many engineering projects to amplify, transmit and control power.

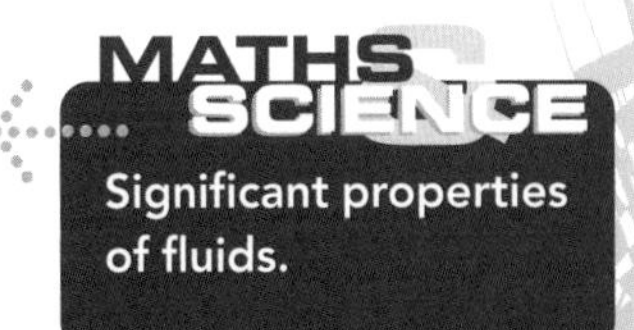


## How fluid power works

In the mid 1600s a scientist called Blair Pascal described how liquids behaved.

Pascal put water into two linked containers, one large and one small.

**FIG 1.2:** Pascal's experiment

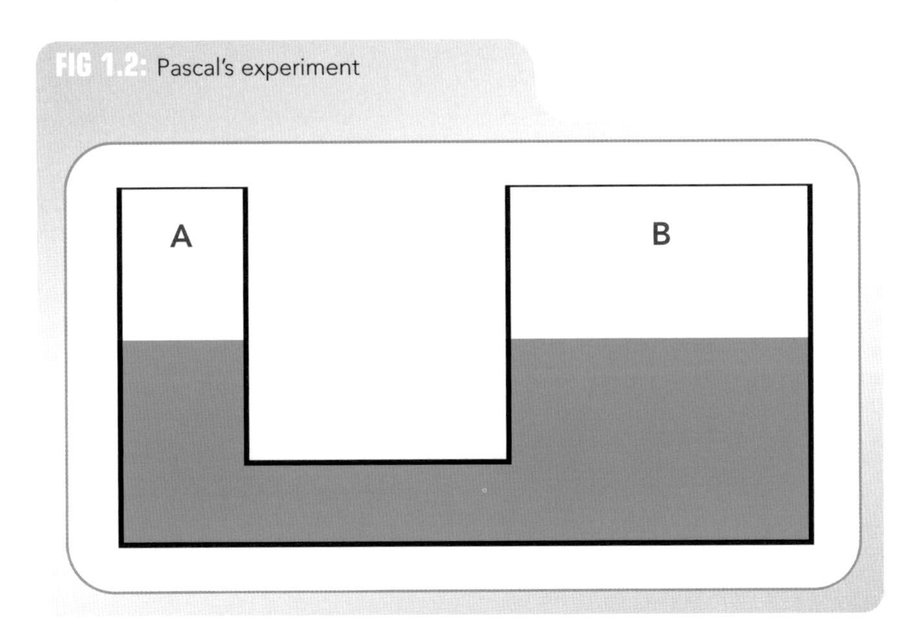


You may think that as there is more water in B the fluid will be under more pressure. If B had more pressure it would push the fluid level of A up; however, both levels stay the same.

As a result of his experiments, Pascal stated that the pressure in a static hydraulic fluid in a closed system is the same everywhere.

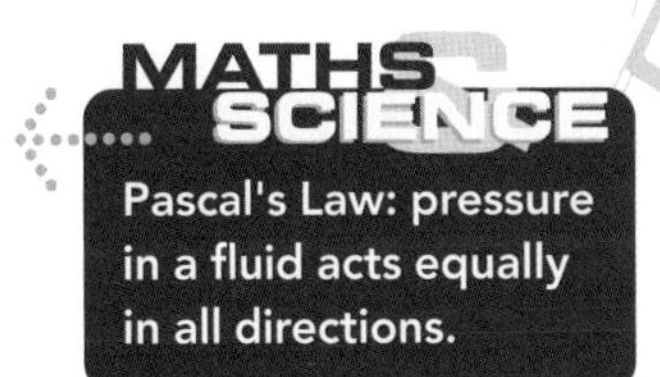


Now let us consider how this applies to hydraulic cylinders and hydraulic systems.

Helipebs designs hydraulic cylinders. A hydraulic cylinder contains a piston which fits snugly into a larger cylinder. The piston moves under fluid pressure.

FIG 1.3: Simple hydraulic system

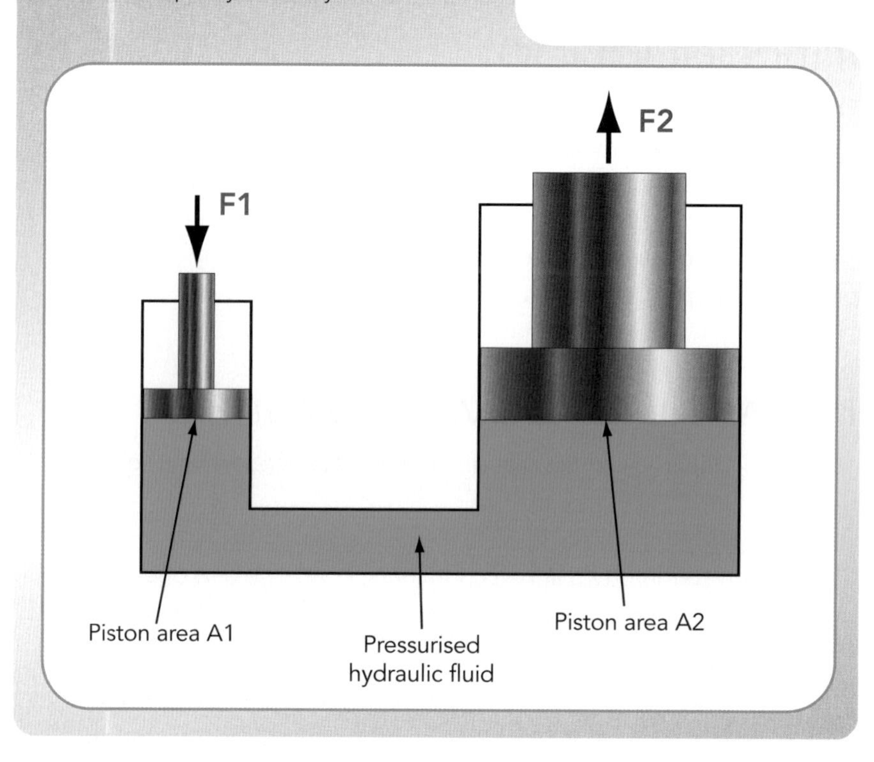


**MATHS & SCIENCE**

An applied force is the pushing or pulling that acts on the outside of an object. The unit of force is newtons (N).

A simple hydraulic system consists of two different-sized cylinders connected by a pipe. Force applied to one point is transferred to another point using a fluid.

So pressure exerted on the smaller piston is transmitted through the fluid to act on the internal surface of the larger piston. As the pressure is constant throughout a closed system (Pascal's Law), it is experienced equally by each piston. However, because each piston has a different surface area, the force exerted on each piston will be different, even though the pressure is the same.

**MATHS & SCIENCE**

In a hydraulic car braking system, why do we have to make the piston down by the wheel bigger than the piston by the brake pedal?

This is because:

**Force = Pressure x Area**

If the larger piston has twice the area of the smaller piston then the force on the larger piston will be twice as great. In order to create that extra force, the smaller piston has to be moved by twice the distance.

This is the principle of hydraulic multiplication. If a piston had ten times the surface area of another piston, when the force is applied to the smaller piston, it will apply ten times the force to the larger piston. To achieve this, the smaller piston will move ten times the distance.

**FIG 1.4:** Demonstration of hydraulic multiplication

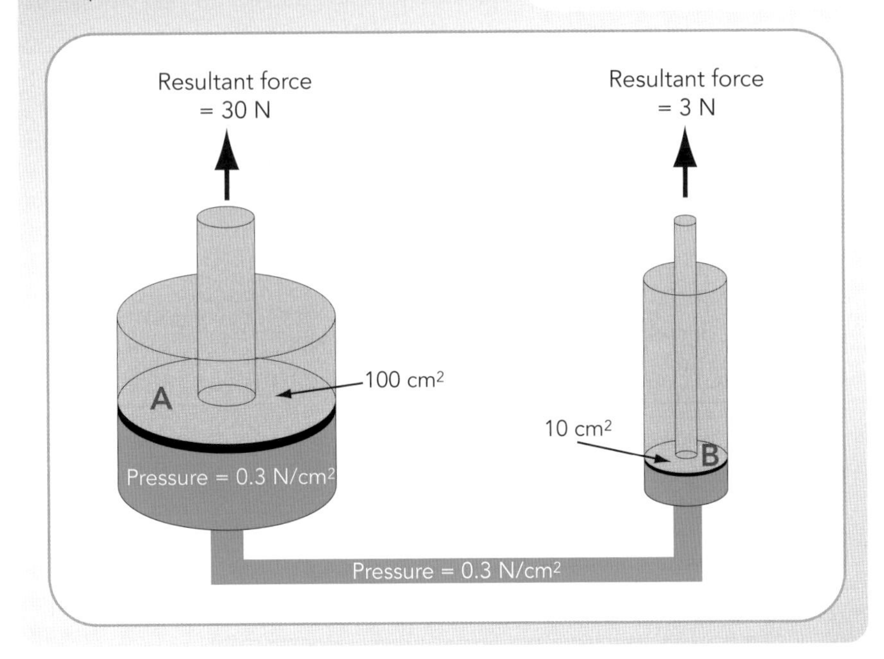


**MATHS & SCIENCE**

**Hydraulic multiplication explains why the small force applied to the brake pedal of a car is turned into the large force needed to stop or slow the car.**

So by changing the size of one piston and cylinder relative to another you can alter the applied force.

You could design a spreadsheet model to explore the size (the area) of cylinder needed to lift an object (the force) for any given fluid pressure.

# The hydraulic cylinder

A **cutaway** of a hydraulic cylinder designed by Helipebs is shown below.

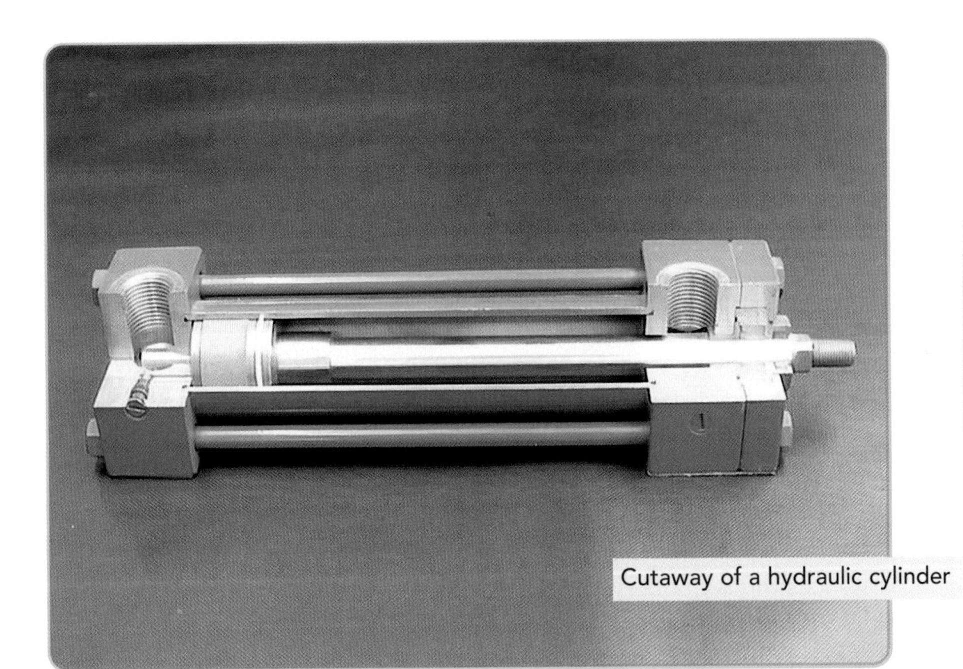

Cutaway of a hydraulic cylinder

**cutaway** – a diagram or model of an object with part of the outer layer removed to reveal the interior

**schematic drawing** – a diagram that represents the elements of a system

As you can see, a hydraulic cylinder consists of six main parts:

- a front head
- a rear head
- tie rods
- piston rod
- cylinder tube
- piston.

A **schematic drawing** of a hydraulic cylinder is shown below.

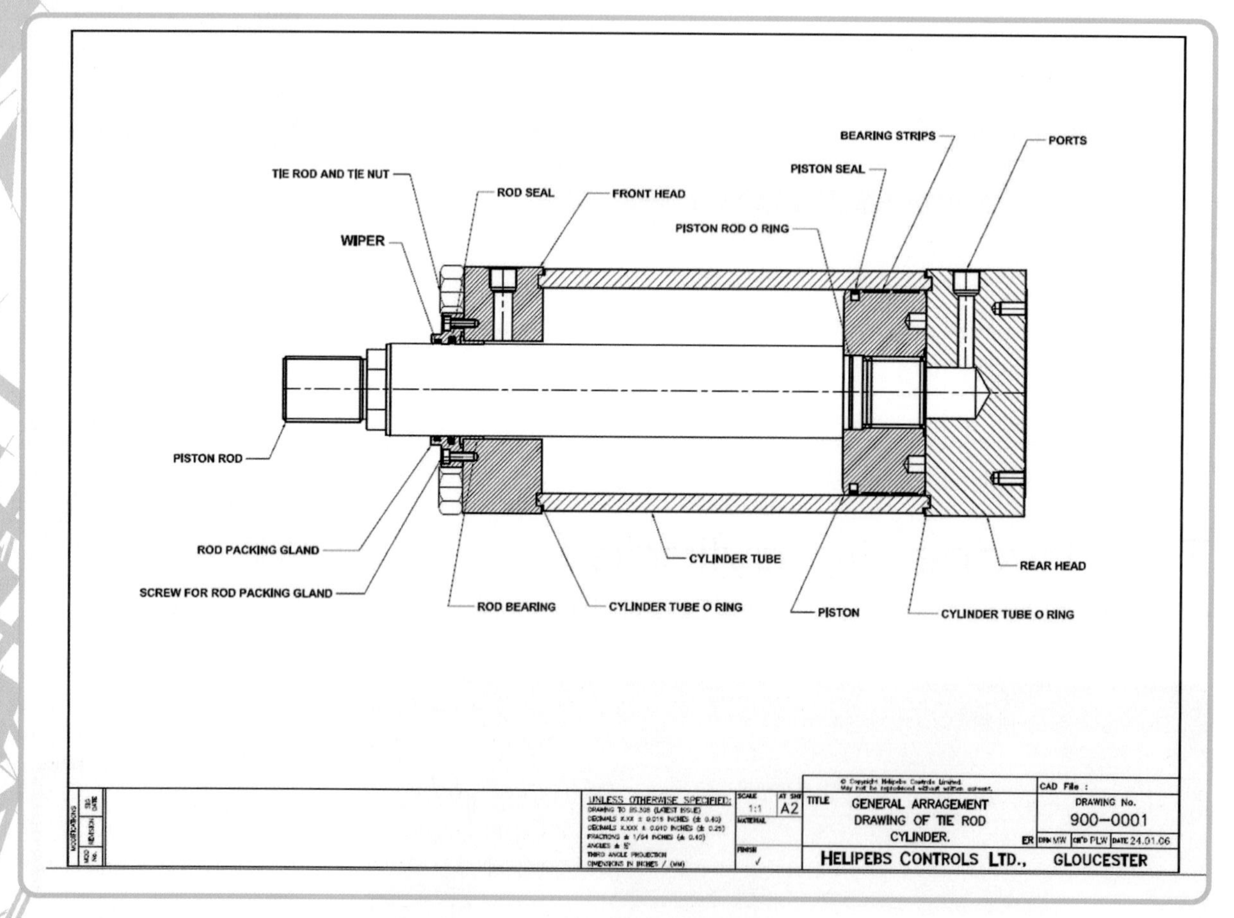


## Critical points

**friction** – force that resists movement between two surfaces which are in contact

To work effectively there must be a tight fit between the two parts of a cylinder to prevent the fluid from leaking. If the fluid leaks, pressure will be lost. The problem is that when any two materials rub together they cause **friction**.

Friction is the force between two surfaces rubbing together. In some instances this is good. If you had no friction between your shoes and the floor you would slip over. Imagine the effect of low friction between car tyres and the road.

If the pushing force is greater than the friction, then the object will slide or move. The friction is called kinetic friction, which means moving friction.

Friction inside a cylinder tube will slow the piston rod down, reduce efficiency and wear out the seals. However, there is less friction when there is a liquid present (e.g. oil) between the two surfaces. As the cylinders designed by Helipebs are driven by hydraulic fluid, this helps greatly reduce the friction.

- When two rough surfaces rub together there is more friction.
- When two smooth surfaces rub together there is very little friction.

Therefore, Helipebs must make the inside surface of the cylinder tube as smooth as possible. However, if you make a cylinder tube and piston very flat and smooth and remove all surface particles, the smooth flat surfaces will actually stick to each other, making what is called a '**cold weld**'. So care must be taken in choosing the correct materials.

To help choose materials, an engineer will refer to the materials' **coefficient of friction**. This is the ratio of the force causing a body to slide along a plane and the force normal to a plane.

The lower the number, the easier it is to slide one material over the other. The chart in Figure 1.5 shows the coefficients of friction for a wide range of materials.

Look at the chart in Figure 1.5 on the next page and see if you can explain why brake pads on a car are often made of brake material inside a cast iron drum. But also note how the coefficient of friction reduces if the cast iron drum is wet. The same effects can be seen with rubber used for car tyres and asphalt used for roads. Material is being chosen here because it has a high coefficient of friction.

If you look back at the schematic drawing of the hydraulic cylinder, you will notice that Helipebs has also added a bearing strip and scrapers to the piston. These components are called sealing components. They make the cylinder surfaces more slippery and so reduce friction. They are chosen because they have a low coefficient of friction. The smoothness of the cylinder tube and special low-friction material of the piston seal reduce the friction to a minimum.

## Applying hydraulic pressure

In practice, few hydraulic systems use a cylinder to apply the force (pressure) to the hydraulic fluid. Cylinders are normally used as the output of the system. It is normal to use a hydraulic pump to pressurise the fluid. The simplest type of pump is a gear pump, although these can only apply relatively low pressures (140–180 bar).

**MATHS & SCIENCE**

**Friction provides the force to accelerate, stop or change the direction of a car. Ice and water on the road reduce this friction and make it easier to skid.**

**cold weld** – welding of two materials under high pressure or vacuum without the use of heat

**MATHS & SCIENCE**

**A way of reducing friction is to use ball bearings, as the balls allow the surfaces to move easily without actually touching each other.**

**MATHS & SCIENCE**

**Have you ever felt a piece of sandpaper and a smooth piece of silk? In which case is the coefficient of friction greater?**

**MATHS & SCIENCE**

**The SI unit of pressure is the pascal (Pa). For pressure derived from applied force the unit used is bar or psi. 1 bar = 100,000 Pa.**

**FIG 1.5:** Coefficient of friction of various materials

| Material combinations | | Coefficient of friction | |
|---|---|---|---|
| | | Clean and dry surfaces | Lubricated and greasy surfaces |
| **Material One** | **Material Two** | | |
| Aluminium | Aluminium | 1.05–1.35 | 0.3 |
| Aluminium | Mild steel | 0.61 | |
| Brake material | Cast iron | 0.4 | |
| Brake material | Cast iron (wet) | 0.2 | |
| Brass | Steel | 0.35 | 0.19 |
| Brass | Cast iron | 0.3 | |
| Bronze | Steel | | 0.16 |
| Bronze | Cast iron | 0.22 | |
| Bronze – sintered | Steel | | 0.13 |
| Cast iron | Cast iron | 1.1, 0.15 | 0.07 |
| Cast iron | Mild steel | 0.4, 0.23 | 0.21, 0.133 |
| Carbon (hard) | Carbon | 0.16 | 0.12–0.14 |
| Carbon | Steel | 0.14 | 0.11–0.14 |
| Copper-lead alloy | Steel | 0.22 | |
| Copper | Copper | 1 | 0.08 |
| Copper | Cast iron | 1.05, 0.29 | |
| Copper | Mild steel | 0.53, 0.36 | 0.18 |
| Glass | Metal | 0.5 - 0.7 | 0.2–0.3 |
| Glass | Nickel | 0.78 | 0.56 |
| Graphite | Steel | 0.1 | 0.1 |
| Iron | Iron | 1 | 0.15–0.20 |
| Lead | Cast iron | 0.43 | |
| Magnesium | Magnesium | 0.6 | 0.08 |
| Nickel | Mild steel | 0.64 | 0.178 |
| Nylon | Nylon | 0.15–0.25 | |
| Phosphor-bronze | Steel | 0.35 | |
| Platinum | Platinum | 1.2 | 0.25 |
| Polythene | Steel | 0.2 | 0.2 |
| Rubber | Dry asphalt | 0.5–0.8 | |
| Rubber | Wet asphalt | 0.25–0.75 | |
| Rubber | Dry concrete | 0.6–0.85 | |
| Rubber | Wet concrete | 0.45–0.75 | |
| Steel | Steel | 0.8 | 0.16 |
| Teflon | Steel | 0.04 | 0.04 |
| Tungsten carbide | Steel | 0.4–0.6 | 0.1–0.2 |
| Tungsten carbide | Tungsten carbide | 0.2–0.25 | 0.12 |
| Tungsten carbide | Copper | 0.35 | |
| Tungsten carbide | Iron | 0.8 | |

A gear pump works by rotating gears sucking hydraulic fluid in and forcing it out to the output part of the pump. The teeth in the gears **mesh** to prevent the hydraulic fluid from flowing back.

**mesh** – be engaged with another gearwheel
**cam** – shaped piece of metal or plastic fixed to a rotating shaft
**eccentric** – not placed centrally
**compress** – to press something into a smaller space

FIG 1.6: Gear pump

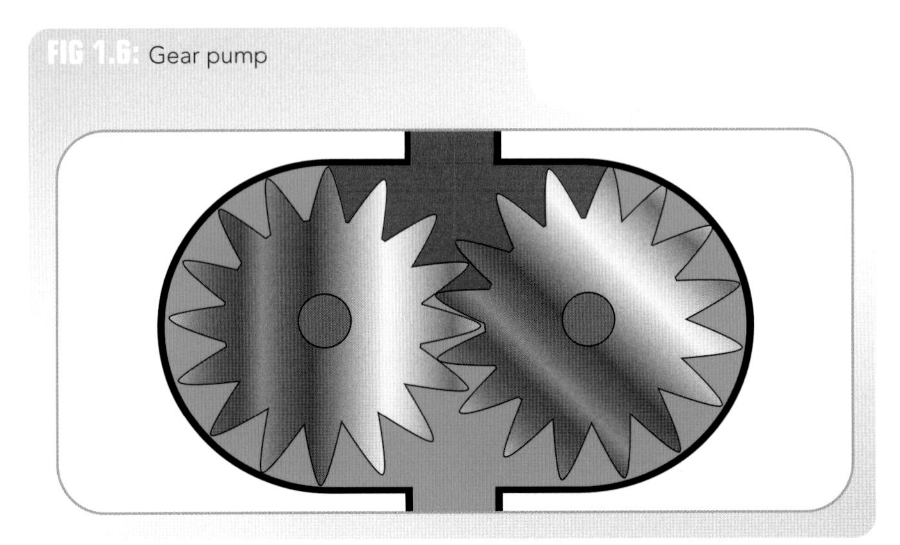

## Piston pump

A piston pump works by using a **cam** to push pistons in, forcing the hydraulic fluid along a pipe and through a one-way flow valve.

FIG 1.7: Piston pump

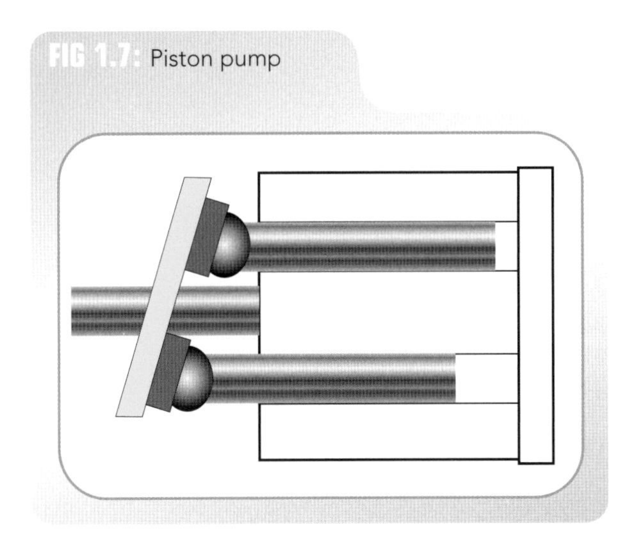

FIG 1.8: Vane pump

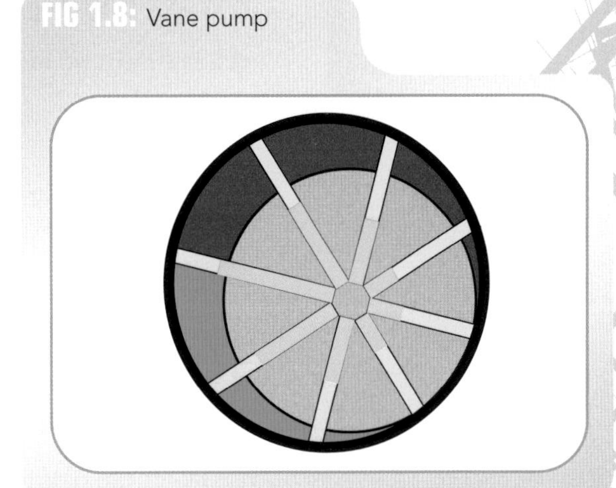

## Vane pump

A vane pump can achieve pressures of 200 bar. The biggest advantage of a vane pump is its vibration-free operation. Vane pumps are also very quiet compared to other pumps. Vane pumps work using an **eccentric** rotating cylinder, which collects the hydraulic fluid and then **compresses** it against the outer wall.

**To find out more about pumps go to www.pump school.com**

# The process

## Planning

### The brief

Helipebs receives orders from customers in the form of a **design brief**. The original customer brief would normally come via telephone, email, fax, website or external salesman. The brief could come in the form of a sketch with words, or just words. If it is a specific complicated design, the brief could be very long. Recently the company received a customer brief of 97 pages.

Often, however, customer briefs do not contain all the information that is required, as the example brief, below, shows.

Example customer brief:

- 2 of cylinders 10″ bore x 4.5″ rod x 8.5″ stroke.
- Cylinders to be able to work at a sea depth of 6562 feet (2000 metres).
- Work on **biodegradable** mineral oil.
- Seals must be resilient to methanol.
- Painted to customer's specification number: MJW 8.21.60 Revision D.

Before proceeding with a design, further questions need to be asked to clarify the following:

1. The working pressure of the cylinders (very important).
2. The temperature range the cylinders will be working in.
3. The speed of operation.
4. The length of time the cylinders will be operating in subsea conditions.
5. The piston rod extension from the front face of the cylinder, and what end connection is required, i.e. size of thread, male/female and how long/deep.
6. The mounting style, i.e. front/rear flange, trunnion, spherical eye, etc.

All of the above would have to be resolved before the design is completed. This could be done by several different channels, i.e. through the sales department, or the quickest way: by the engineer talking to/emailing the customer's engineer.

**design brief** – document that details design requirements, such as function, performance, size, materials, etc.

**biodegradable** – able to decay naturally and harmlessly

MATHS & SCIENCE

1 foot is about 30 cm. 1 inch is about 2.5 cm. The symbol for inch is ″.

## The design

Hydraulic cylinders are quite difficult to design. The designer has to consider the force needed, the distance travelled and the working conditions. Helipebs designs cylinders that work under the sea. This adds not only external pressure and corrosion considerations, but also environmental considerations.

The design considerations, however, are only the start. Choosing materials that can safely withstand the extreme pressures is vital. Service and maintenance also need to be considered, as does cost.

The solution would involve designing the cylinder and producing an **installation drawing** showing all the external dimensions of the cylinder. However, you can only produce this after you have fully drawn up a **general arrangement drawing** showing how all individual parts fit with each other.

**installation drawing** – provides instructions on the installation requirements

**sourced** – to get something from a particular place

**design parameters** – the design limits which define how something can be done

**design envelope** – summarises design requirements, constraints and criteria

**acceptance of order** – the action of consenting to undertake the specific client design requirements

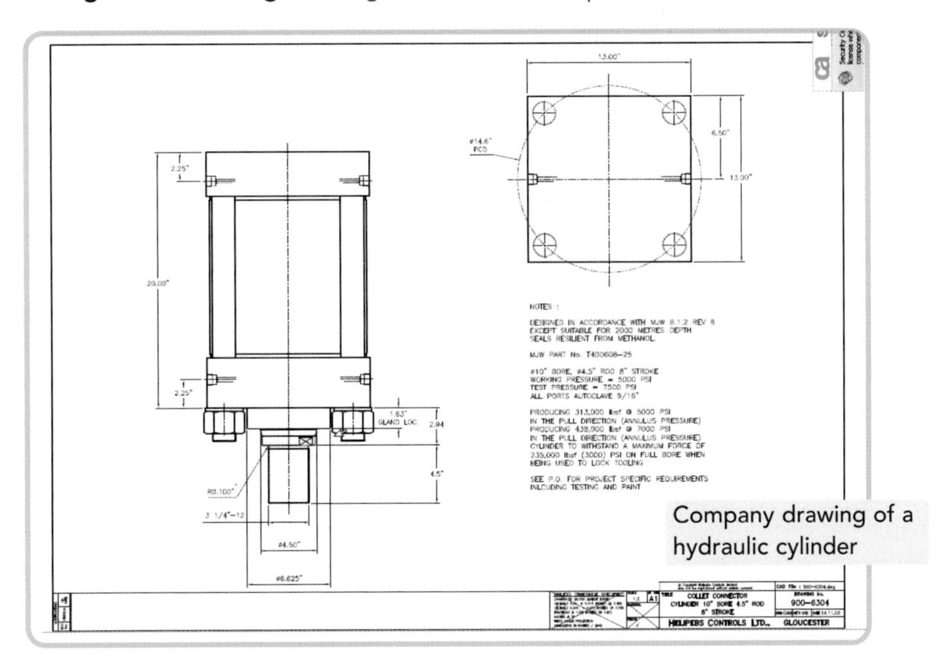

Company drawing of a hydraulic cylinder

## Final steps

Materials would be **sourced** and calculations performed to ensure that the proposed design is capable of withstanding the internal/external pressures and any other **design parameters**. Any health and safety issues would be addressed at this time. This allows the customer to see if the proposed design meets their requirements and to see if it fits the **design envelope**.

The installation drawing is the final specification. It will contain all the relevant specification details plus the actual drawing of the cylinder. An **acceptance of order** is sent to the customer for approval, but it is just an acknowledgement of the customer's purchase order.

When the customer has approved the drawing, a formal quote is raised and sent to them. This would include the price and a delivery date.

HEALTH & SAFETY

If fluid leaks occur, there is the potential of a fire hazard. Some hydraulic machines require fire-resistant fluids, depending on their applications.

The detailed drawings would only start when the order has been received.

The final stage is to finish the design/details in the most cost-effective way. This involves the engineer knowing the product and the capabilities of the production workshop.

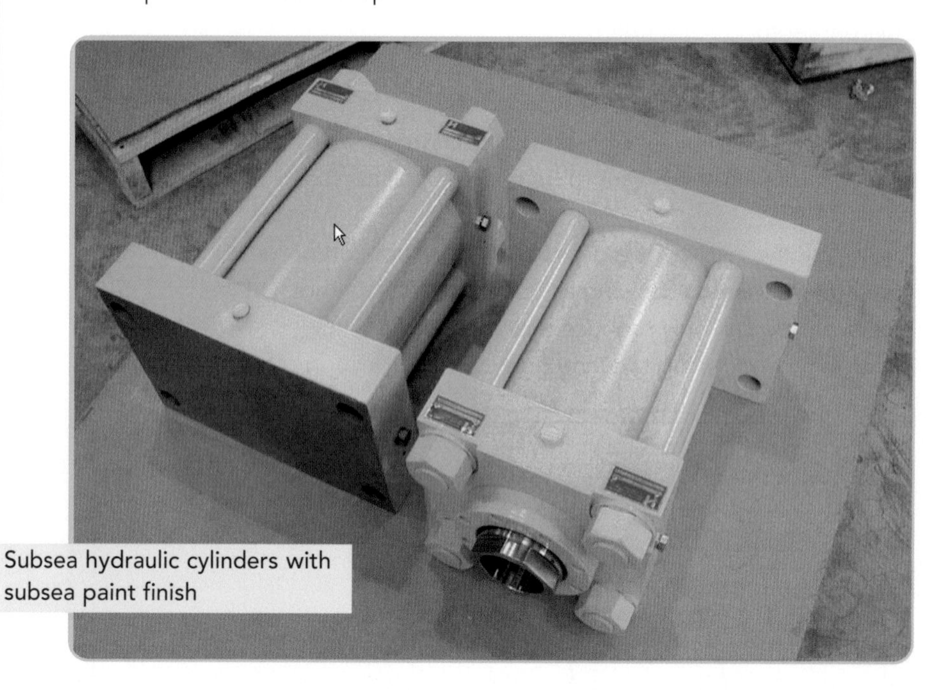

Subsea hydraulic cylinders with subsea paint finish

## Production

### Methods and equipment

Helipebs uses a wide range of traditional manufacturing methods. This is due to the '**one-off**' nature of their products. The term used in the film is **bespoke**, meaning they design products for the customer's specific needs.

**one-off** – something that is made only once
**bespoke** – made to order

The list below shows the machines used in the workshop to make tie rod cylinders. Different machines are used depending on the size of cylinder and the quantities involved.

- Colchester Mascot 1600 x 3 lathe
- Tie rod rolling machine x 2
- Crowthorn lathe
- Herbert capstan lathe x 3
- Colchester Mascot 1400 long bed lathe
- Lapping wheel machine x 2
- Hand grinding machine x 3
- Linisher machine vertical x 3
- Surface grinder
- Various radial drills x 6

- Dean & Smith Grace long bed lathe
- Colchester Triumph 2000 lathe
- Leadwell milling/drilling centre x 2 (CNC)
- Index GU1000 CNC turret lathe
- GH-1660ZX centre lathe
- Milling machine x 3
- Various saws x 3
- Omega 3 jig borer
- Leadwell CNC turret lathe
- Colchester CNC 4000 turret lathe
- SL-3 Mori Seike CNC turret lathe x 2

Many of these machines will be similar to those used in a school or college workshop.

Machining operation on a lathe

The **turret lathe**, also known as a **capstan lathe**, is a form of metal cutting lathe that is used for short production runs of bespoke parts.

A capstan lathe uses **turning** techniques to remove metal. Turning is used to produce **solids of revolution**. The process can be **tightly toleranced** because of the specialised nature of the operation. In most turning operations the tool is stationary and the part is rotated. The workpiece is mounted on the **chuck**, which rotates relative to the tool.

In the film you will see two metal-removal operations, turning and **facing**.

**capstan lathe** – has a revolving turret that holds several tools
**turning** – the lathe can be used to reduce the diameter of a part
**solids of revolution** – solid figure obtained by rotating a plane figure around some straight line that lies on the same plane
**tightly toleranced** – strict allowance made for something to deviate in size from a standard
**chuck** – a device for holding an object firmly in a machine
**facing** – the process of removing metal from the end of a workpiece to produce a flat surface

Colchester CNC-4000

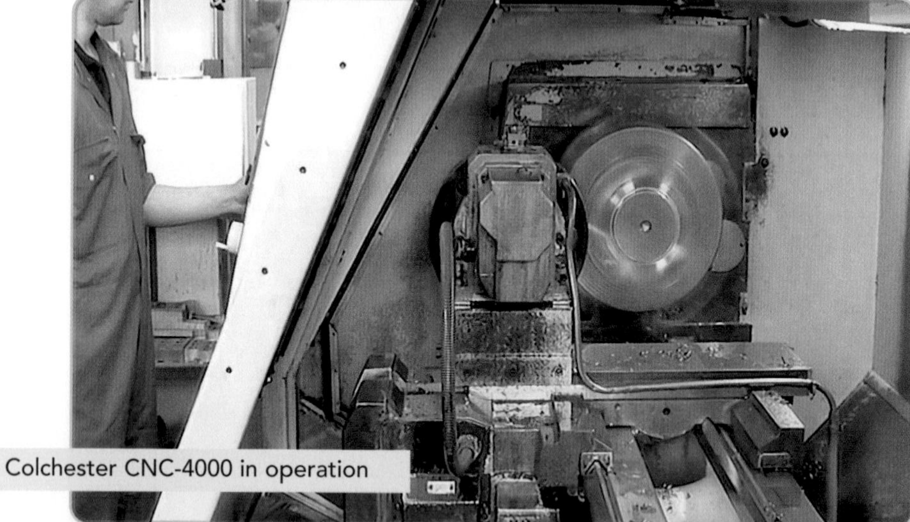
Colchester CNC-4000 in operation

**NUTS & BOLTS**

To find out more about lathes and their operation go to www.technologystudent.com

Close-up view of an operation on Colchester CNC-4000

Facing is the name given to describe removal of material from the flat end of a cylindrical part. Facing is often used to improve the finish of surfaces that have been parted.

## Materials

The types of materials used by Helipebs are listed below.

- All types of carbon steels, i.e. low-strength mild steel up to high-tensile hardening steels. This would include chrome-plated round bar.
- Different types of stainless steel, including chromed bar, e.g. BS 970-304,316L, 416S29 and 17/4 PH.
- Inconal 718 and 725.
- Monel K400 and K500.
- Nickel aluminium bronze.
- Aluminium.
- Bronze.
- Aluminium bronze.

## Special fluids

The fluids listed below are all used at Helipebs in their hydraulic cylinders. Many of these fluids have special characteristics that make them useful in particular applications.

- Castrol Transaqua HT: water-glycol mixture.
  Used by Helipebs in test rigs for subsea cylinders.
- Castrol Brayco Micronic 864: water-glycol mixture.
  Used by Helipebs in test rigs for subsea cylinders.
- Houghto-Safe 620: water-glycol mixture.
  Used by Helipebs in test rigs for subsea cylinders.
- Castrol AWS 32: standard mineral oil.
  Used by Helipebs in test rigs for all standard-type cylinders.

## Tolerance

**Tolerance** in engineering is the defined range of acceptable variation in a dimension or value. For a hydraulic cylinder, this is the dimensions and **parameters** that may vary within certain limits without significantly affecting the effectiveness of the cylinder.

Tolerances are specified to allow room for **imperfections** in the manufacturing of parts and components without compromising the performance of the cylinder.

It is good engineering practice to specify the largest possible tolerance while maintaining proper functionality. Closer or tighter tolerances are more difficult and costly to achieve and are likely to lead to more rejects. On the other hand, larger or looser tolerances may significantly affect the operation of the cylinder.

**tolerance** – the amount by which a measurement or calculation might vary and still be acceptable

**parameters** – a set of facts or a fixed limit which establishes or limits how something can or must be done

**imperfections** – a fault, blemish or undesirable feature

**MATHS & SCIENCE**

psi stands for pounds per square inch. 1 psi = 6894.76 Pa.

**HEALTH & SAFETY**

Pressurising a cylinder rated at 5000 psi to 7500 psi can have devastating consequences. Failure of the hydraulic cylinder can result in an explosion.

**certificate of conformity** – a certificate to show that the product has been assessed and conforms with stated requirements

**NUTS & BOLTS**

To find out more about product testing and certification go to www.bsi-global.com

## Testing

Testing of hydraulic components is a vital part of the process. In the film you will see a cylinder being tested. A cylinder required to work at a pressure of 5000 psi (344 bar) is actually tested to a pressure of 7500 psi (517 bar) to test its integrity. To put these values into context for you, a car tyre works at a pressure of 35 psi and a bike tyre at about 65 psi.

The hydraulic cylinders are tested in batches and, overall, the test will take approximately eight hours. During the test the whole testing area is sealed off due to the potential dangers involved.

A **certificate of conformity** is produced for each cylinder.

FIG 1.9: Certificate of conformity

HELIPEBS CONTROLS
Sisson Road
Gloucester
GL2 0RE
Telephone : 01452 423201
Fax : 01452 307665
Website : www.helipebs.co.uk

CERTIFICATE OF CONFORMITY

CUSTOMER :- MJW ENG LTD. T.I.P. No. :- ..........

CUSTOMER PART No. :- **T400608-25** ORDER No. :- MJW4523

DESCRIPTION OF GOODS :- 10" x 4.5" x 8" Hyd Cylinder.

CONTRACT No. :- G0814

SERIAL No. :- 092356 / 092357

H.C.L. DRG. No. :- 900-6304..........

CLEANING PROCEDURE:- BF-003 Issue B to NES341 Pt 1 Grade D

ASSEMBLED BY:- .......... No. :- A .......... DATE :- ..........

| TEST (s) | COMMENTS |
|---|---|
| .......... | Test Pressure 515 Bar. |
| .......... | .......... |
| .......... | .......... |
| .......... | .......... |
| .......... | .......... |

TESTED BY :- .......... No. :- Q .......... DATE :- ..........

WE HEREBY CERTIFY THAT THE MANUFACTURE DETAILED HEREON HAVE BEEN TESTED / INSPECTED IN ACCORDANCE WITH THE CONDITIONS AND REQUIREMENTS OF THE PURCHASE ORDER, AND UNLESS NOTED ABOVE, CONFORM IN ALL RESPECTS TO THE SPECIFICATIONS, DRAWING(S) RELEVANT THERETO.

FINAL INSPECTION :- .......... DATE :- ..........

HEALTH AND SAFETY DECLARATION

HELIPEBS CONTROLS LTD, recognise the statutory obligation placed upon the company. We have taken all reasonable precautions to provide products which are safe and without risk to health when correctly utilised.
Prior to installation and / or putting the product into service you should ensure that it fully complies with the requirements of your purchase order and also that the parameters indicated above are adequate to meet the duty to which to you are committing the product.
We emphasise the necessity to follow all advice available concerning the correct use of any products supplied by HELIPEBS CONTROLS, whether it concerns the application or the Health & Safety precautions connected with the product(s). If you should have concern over any aspect of safety then please contact our Engineering or Technical Sales Department(s), who will be pleased to assist you.

HELIPEBS CONTROLS LTD. QD. 026

# Critical points

When you watch the film you will notice a number of signs in the factory. Many of these are health and safety related, but others relate to the quality standards of the company.

One sign reads: Rejects cost money and customers.

Another reads: Quality comes as standard.

Company signage to promote quality

All manufacturing companies have to be aware of how 'rejects' (products or components that fail to meet the necessary tolerance) cost money, not only in terms of the materials and processes used to reach this stage but also in terms of time delays in delivery to the customer.

In the film you will hear Glen talk about the pressures of time. Any delay in the delivery of a product can cost the customer large amounts of money, as the hydraulic cylinders are a small part of a much larger project. He talks about the problem of having large numbers of fitters on an oil rig standing around waiting if delivery schedules are not met and a crucial piece of equipment (the hydraulic cylinder) fails to arrive.

**penalty clauses** – found in contract agreements and provides for a penalty in the event of default

Companies often add **penalty clauses** to the contract when they agree to buy products, which mean that the company producing the product has to carry the cost of any delays caused by them.

Let us now explore the costs involved in any reject.

- The first cost is the cost of the materials. As Helipebs uses high-quality materials, this can be quite high.
- The second cost involves the transportation of these materials to each stage of manufacture.
- The third cost is the machining/manufacturing costs up to the point of product rejection. This includes the cost of the engineer, the cost of heating and lighting, the cost of storage and the cost of wear and tear of machine tools and machinery.
- Lastly we have the cost of the time delay in delivery to the customer, and, of course, the fact that the customer will probably go to another supplier next time, costing the business future orders and its reputation. Remember the sign 'Rejects cost money and customers'.

As can be seen, the cost of rejects to a manufacturing company is high. As such, Helipebs uses detailed mapping and quality assurance techniques throughout the whole factory to ensure high levels of quality product. Everyone is involved in quality; as the sign says, 'Quality comes as standard'.

# The career path

## PERSONAL PROFILE: GLEN HARDING

### Journey to a design engineer

I have always had an interest in how things work, to the extent of taking everything apart that showed the slightest sign of a fault.

Progressing through senior school I developed an interest in art, which continued to college. However, after six months I became disillusioned and frustrated, and left. I sought a different occupation and decided on a shortlist of engineering, IT and graphic design.

Having landed a position at college on an engineering course, with the intention of going to university, the class was approached by a local company looking for apprentices. This seemed like an ideal situation of learning and earning, so I put myself forward and have never looked back.

**Qualifications**

Hydraulics fluid power (FP 1), (FP 2), (FP 3) – 2001. (FP 4) – 2006.
Higher national diploma in engineering (HND) – 2002.
Higher national certificate in engineering (HNC) – 2001.
National vocational qualification in engineering level 3 (NVQ 3) – 2000.
Engineering modern apprenticeship – 2000.
National certificate in engineering (ONC) – 1999.
National vocational qualification in engineering level 2 (NVQ 2) – 1998.

**Professional experience**

*Design Engineer* *Oct 2003–present* *Helipebs Controls*

- 2D & 3D design work operating AutoCAD & Solid Works.
- Special purpose linear actuators, servo control, subsea applications for oil and gas companies, including collet connector retrieval cylinders.
- Repair, overhaul and design development of linear actuators.
- Design of test frames and fixtures for aerospace and Formula 1 industries.
- Project management from customer liaison and scheduling, through to build and test.
- Stress design calculations using Microsoft Excel.

*Design Engineer* *1997–2003* *Forward Industries*

- 2D & 3D design work operating various CAD software.
- Hydraulic and mechanical systems design work.
- Remote handling equipment, in cell systems work, refuelling and decommissioning tooling for BNFL and RR.
- Development of design on VNE master-slave manipulator arms.
- Hydraulic and mechanical systems design calculations using Microsoft Excel and Lotus.
- Research and development role incorporating skills such as turning, milling and bench work, progressing products from the design stage into manufacture.
- Development testing, preparation of technical specifications and test reports.
- Site work maintenance and installation.
- Hands-on testing of prototype cylinder designs, fitting work when required on complex builds.

**To find out more about a career in the fluid power sector go to www.prospects.ac.uk and www.learndirect-advice.co.uk/helpwithyourcareer/**

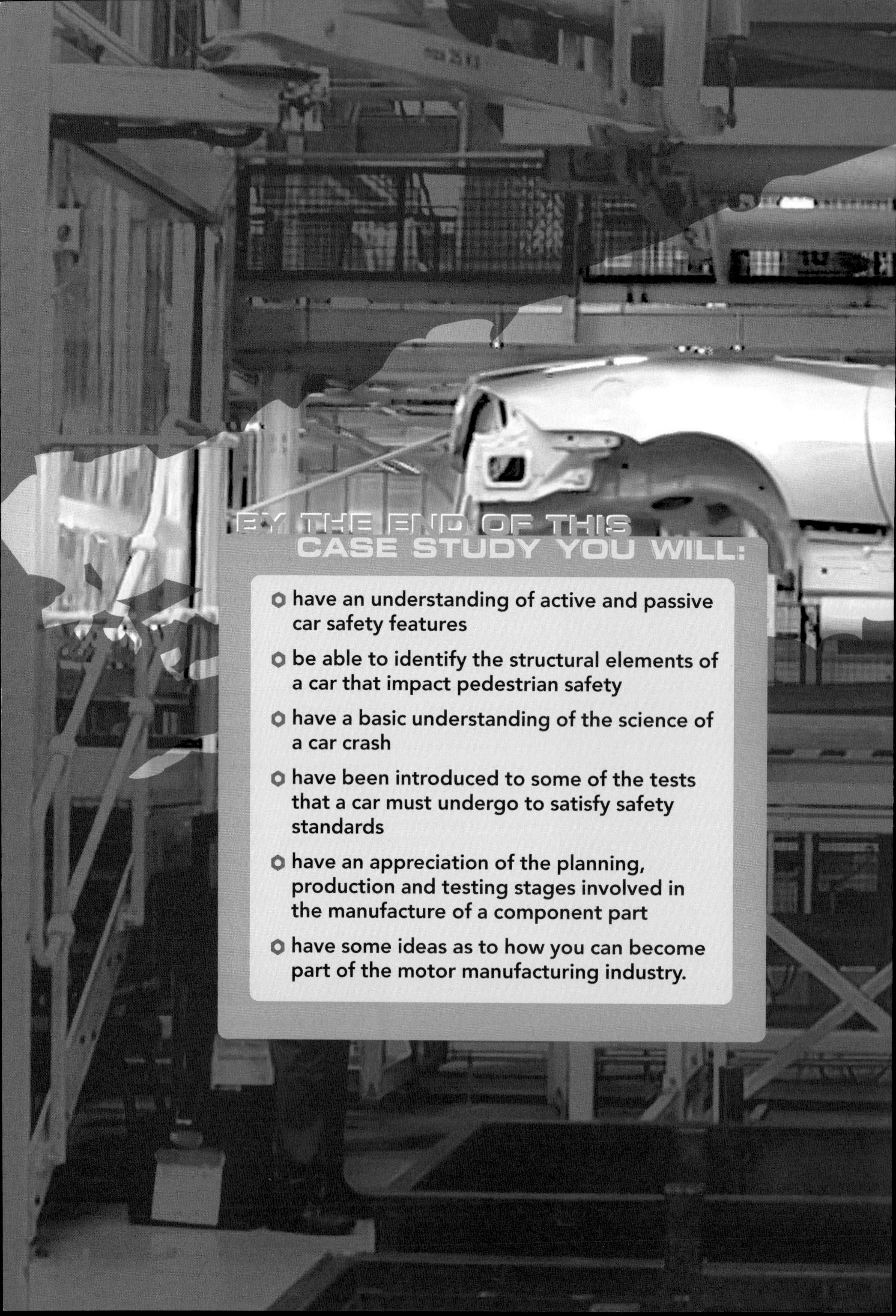
BY THE END OF THIS
CASE STUDY YOU WILL:
have an understanding of active and passive car safety features
be able to identify the structural elements of a car that impact pedestrian safety
have a basic understanding of the science of a car crash
have been introduced to some of the tests that a car must undergo to satisfy safety standards
have an appreciation of the planning, production and testing stages involved in the manufacture of a component part
have some ideas as to how you can become part of the motor manufacturing industry.

# THE MOTOR INDUSTRY: JAGUAR CARS

## Case study overview

This case study introduces you to the motor manufacturing industry and its role in developing cutting-edge safety technology.

Over the following pages you will become familiar with a real company that has developed the first pedestrian protection pop-up bonnet system.

You will consider what happens in a car crash and the scale of the risks to pedestrians inherent in a pedestrian-car collision. This will help you to understand the purpose of the pop-up bonnet system and enable you to appreciate the significance of new safety technology.

Finally, you will be introduced to a technical specialist from the company who will describe the route he took to his current position.

# The sector

## The motor industry

The motor industry, also known as the motor manufacturing industry, comprises companies that design and manufacture cars. The major companies have worldwide bases in countries such as France, Germany, Italy, Japan, South Korea, the United States and the United Kingdom (UK).

In 2006, the **annual turnover** of the UK retail motor industry was in excess of £60 billion and it employed almost 600,000 people.

In 2005, 1.6 million cars were produced in the UK alone.

NUTS & BOLTS

To find out more about the motor industry go to www.motor.org.uk

**annual turnover** – the amount of money taken by a business in a year
**discipline** – a particular area of study
**aesthetic** – concerned with beauty

## Car design

The design of modern cars is carried out by a large team of designers and engineers from many different **disciplines**. These include **aesthetic** designers, production engineers, electrical engineers, manufacturing engineers, safety engineers and many other specialists, often employed in small companies that supply the motor manufacturing industry. These small companies will often focus on areas such as seat design, speedometers, tyres and a wide range of similar components. At the heart of all of these vital components is an understanding of both scientific principles and manufacturing methods.

As part of the product development cycle, teams of designers will work closely with teams of design engineers responsible for all aspects of the vehicle. These engineering teams include: chassis, body, safety, electrical and production engineering. The design team will often have both an exterior designer (responsible for the design of the exterior of the vehicle, such as the proportions, shape and surfaces of the vehicle) and an interior designer (responsible for the design of the vehicle interior, such as the proportion, shape and surfaces for the seats, trim panels, etc.).

## Car safety

Road traffic injuries represent about 25 per cent of worldwide injury-related deaths with an estimated 1.2 million deaths each year. Over 8,400 pedestrians and cyclists die on European Union (EU) roads annually and over 170,000 are seriously injured. Most are hit by the fronts of cars in urban and residential areas, and the majority of these are children and elderly road users. In EU countries, pedestrians have a nine times higher death risk than car occupants. In several European countries there are annual increases in the number of pedestrian and cyclist deaths.

Despite the scale of the problem, most previous attempts at reducing pedestrian deaths have focused solely on education and traffic regulations. However, in recent years, car design engineers have begun to use design principles that have proved successful in protecting car occupants to develop vehicle design concepts that reduce the likelihood of injuries to pedestrians in the event of a car-pedestrian collision. These involve redesigning the bumper, bonnet and windshield to be energy-absorbing (softer) without compromising the **structural integrity** of the car.

**structural integrity** – a measure of the quality of construction and the ability of the structure to function as required

It is the car design engineers' greatest challenge: incorporating cutting-edge safety technology whilst retaining the car's appeal to customers.

# The company

## Background

Few makes of car on the road today have a heritage as rich and distinguished as that of Jaguar. Jaguar was conceived in 1922 by William Lyons, a man whose original vision was to design motorcycle sidecars with his Swallow Sidecar Company.

By 1927, he had progressed to building special-bodied cars, which in 1931 resulted in the launch of the legendary SS1 – a car which set the stage for the first true Jaguar. As the range improved and expanded, it needed a name to reflect its speed, power and sleekness, and in 1935 the Jaguar name was born.

In the mid-fifties, Jaguar had reached a point in its history of selling only luxury and sports vehicles. The company also sold a great deal of its production in foreign markets. Jaguar needed to cement a stronger position by producing a car that could be sold in their UK home market and to a larger market. Thus, the Jaguar MK I was introduced at the 1955 Motor Show.

By the 1960s, Jaguar needed to make another quantum leap forward. The E-type, announced in 1961, was just that. A true automotive icon, and arguably the most famous sports car of all time, some 70,000 Jaguar E-types were built over the next 13 years – with around 60 per cent being shipped to the United States.

In 1968, the XJ6 arrived. It was without question the finest Jaguar saloon yet, and met with instant praise. First and foremost, the shape was another Lyons masterpiece. In an era when cars were starting to lose their character, the Jaguar strongly retained its identity.

The 1980s saw Jaguar continuing to raise the bar in performance with the launch of the XJ-S HE and a true world supercar, the XJ220.

By 1989, the Jaguar board recognised the potential value of collaborating with a world-class car manufacturer. Events moved swiftly, and by the end of the year the company was owned by Ford Motor Company Limited.

Jaguar XK

**NUTS & BOLTS**

**To find out more about Jaguar Cars go to www.jaguar.co.uk**

## Modern day

Throughout the nineties, Jaguar underwent a programme of modernisation and expansion that saw the introduction of the new XK8 and XKR sports coupés and convertibles and the launch of the new mid-sized S-type sports saloon – ensuring that Jaguar would enter the new millennium with record levels of production and the broadest product range in the company's history.

In recent years, Jaguar has made an effort to broaden its product line with the introduction of lower-priced, entry-luxury vehicles. Today, the Jaguar marque is known for offering cars that boast distinctive styling, sporting performance and luxurious interiors.

In February 2001, the new X-type became the highlight of the Geneva International Motor Show, while in 2002 the all-new XJ, featuring a revolutionary aluminium body construction, was unveiled at the Paris Motor Show to great acclaim. In the 2003 Top Gear Survey the XJ8 came top in terms of quality. Worldwide sales of the all-new XK range of sports cars – the most technically advanced that Jaguar has ever built – began in March 2006.

# The product

## Introduction

In this section, we will explore some of the fundamental scientific principles that need to be considered when we look at motor vehicle design.

In recent years, a number of road safety measures have been introduced. These include the wearing of seatbelts, roadside breathalyser tests, cycling lanes, moving lamp posts away from the edge of roads, and speed cameras, all of which have helped reduce the number of deaths and injuries on our roads.

However, the adverts in glossy motoring magazines often focus on the glamorous side of cars, for example the shape, the acceleration and the speed, etc. Jaguar Cars, however, takes passenger and pedestrian safety very seriously and has invested heavily in attempts to reduce the impact of a car accident on a human being. In order to understand how to achieve this we need to explore what happens in car accidents of various types.

There are two types of car safety features: **active** and **passive**. To make driving safer and prevent accidents from occurring, cars may have the following active safety features: **ABS**, **traction control**, etc.

When an accident is imminent, various passive safety systems – for example seat belts and air bags – work together to minimise damage to the individual involved. Much research has been done using crash test dummies to make modern cars safer than ever. Recently, attention has also been given to cars' design regarding the safety of pedestrians in car-pedestrian collisions.

**ABS** – antilock braking system
**traction control** – a means of electronically reducing the power to the driving wheels, to minimise wheel-spin and maximise traction

HEALTH & SAFETY

**Passive safety features help drivers and passengers stay alive and uninjured in a crash. Active safety features help drivers avoid accidents.**

## The structure of a car

Most drivers are familiar with certain parts of the car, such as the **transmission**, steering and **suspension**, but very few drivers would understand the structural part of the car. In scientific terms a structure is defined as a device that is used to transmit forces through space. For example, the structure in a tall building transmits the force generated by wind into its foundations, which resist the tendency for the building to topple over.

In a car, the **torque** produced by the engine and the forward **thrust** of the wheels during acceleration have to be transmitted through the structural

**transmission** – mechanism by which power is transmitted from an engine to the axle in a vehicle
**suspension** – system of springs and shock absorbers by which a vehicle is supported on its wheels

**torque** – a force that tends to cause rotation
**thrust** – the driving force produced by an engine
**chassis** – the base frame of a car, to which the body, engine, suspension and transmission of a car are fastened

part of the car. In early models the structure was called the **chassis**. The chassis consisted of two beams tied together with cross-beams.

FIG 2.1: Vehicle chassis

Of course, a car is more than a chassis – wheels need to be added, with some sort of steering device, an engine needs to be added and various components to connect the engine to the wheels.

It is also difficult to drive a car whilst standing up, so a driver's seat is the next item to be added. The weather also needs to be considered, as water and engines don't mix very well. To protect the engine an engine compartment is added. To protect the drivers and passengers a **passenger cell** has to be added. It was not until the 1940s and 1950s that passenger cells became part of the structure of the car.

**passenger cell** – the body of the car

FIG 2.2: Passenger cell of a car

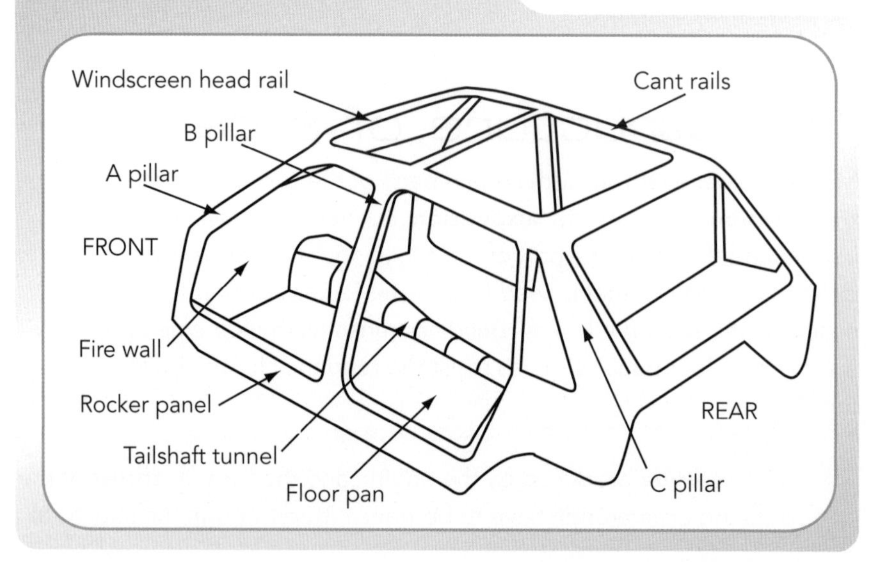

A passenger cell is the technical name given to the cage-like structure that the driver and passengers climb into. Whilst most passenger cells are good at protecting the driver and passengers from the weather, an important consideration is how well the passenger cell protects them in a car crash. The reason that racing car drivers are able to climb out unhurt following an accident is either that their passenger cells have been very carefully designed specifically for this type of crash or, in terms of rally cars, they have been strengthened by adding, for example, **roll bars**.

**roll bars** – metal bars running up the sides and across the top of a vehicle, strengthening its frame

**cross section** – something that has been cut in half so that you can see the inside

The use of the passenger cell as part of the structure of the car is now almost universal. Very few cars have a full chassis any more.

A normal domestic car passenger cell is made from thin steel sheets, pressed into shape and welded together using spot welds. It is worth noting here that, in scientific terms, thin-walled structures do not behave in the same way as conventional structures.

If we compare a simple mild steel beam of rectangular **cross section** with a thin-channelled structure, we can compare the effects on each. The diagram below shows the two beams each loaded with the same maximum load, in terms of the vertical deflection at the point of the load.

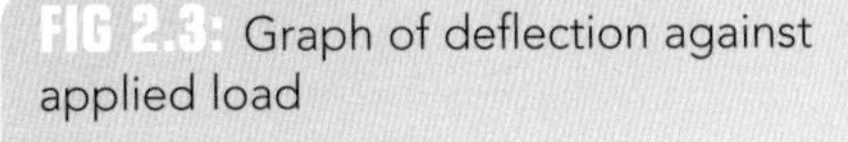

FIG 2.3: Graph of deflection against applied load

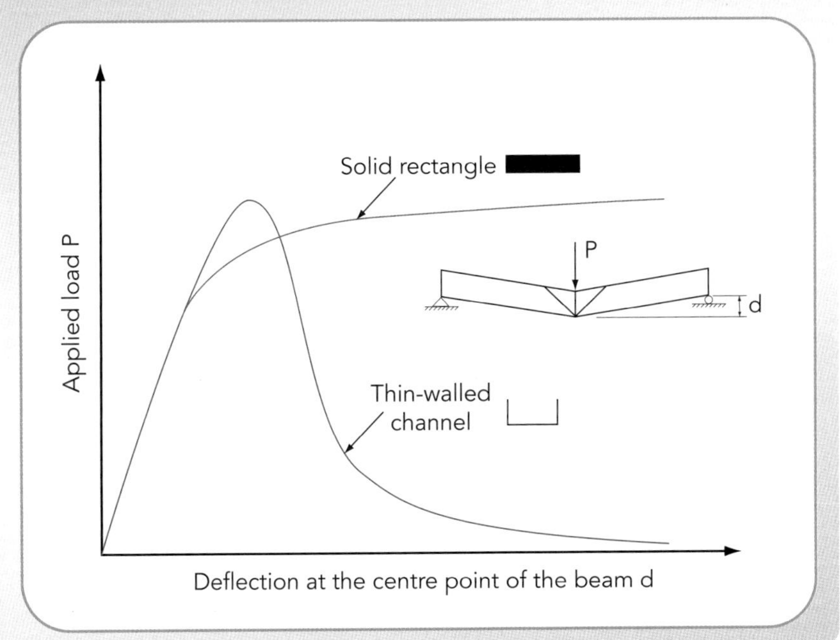


**MATHS & SCIENCE**

**In engineering, deflection is a term that is used to describe the degree to which a structural element is displaced under a load.**

If you look at the graph, you will notice that at first both structures behave very similarly and that the deflection of the structures is minimal. However, if you keep adding load, eventually the rectangular beam will bend over and become almost horizontal, whereas the thin-walled channel will suddenly collapse.

You can see this effect yourself if you squeeze an empty soft drinks can. The circular structure of the can will at first resist your pressure, but then all of a sudden it will collapse, never to return to its original shape.

If you were to examine both structures after the test, you would notice that the cross section of the rectangular section of the beam will have remained unchanged, whereas the channelled section will have changed considerably.

MATHS & SCIENCE

**Elastic deformation is reversible. Once the forces are no longer applied, the object returns to its original shape. Plastic deformation is not reversible.**

As the steel in the thin-channelled structure resists and then permanently deforms, the resistance is said to be plastic.

The depth of the cross section is vital here, as a reduction in the depth will affect both stiffness and strength, important safety elements for a passenger cell.

One of the problems with thin-walled sections is that they collapse suddenly with little or no advance warning. This behaviour is known as brittle behaviour, as it can also arise from the brittleness of the materials being used. Brittleness is a **mechanical property**.

**mechanical property** – the property of a material used as a measurement of how it behaves under a load
**strength** – the tensile strength of a material is its ability to support a load without breaking
**hardness** – the property of a material to resist dents, scratches and cuts
**elasticity** – the ability of a material to return to its original shape after a load is removed
**plasticity** – the quality of being easily shaped or moulded

# Critical points

## Mechanical properties

**Strength**, **hardness**, toughness, **elasticity**, **plasticity**, brittleness, malleability, ductility and corrosion resistance are all mechanical properties used as measurements of how metals behave under a load. These properties are described in terms of the types of force or stress that the metal can withstand and how these are resisted. The chart below shows a range of materials in order of their main mechanical properties. The materials have been listed in descending order of having the property given in the column heading.

| Toughness | Brittleness | Malleability | Ductility | Corrosion resistance |
|---|---|---|---|---|
| Copper | White cast iron | Gold | Gold | Gold |
| Nickel | Grey cast iron | Silver | Silver | Platinum |
| Iron | Hardened steel | Aluminium | Platinum | Silver |
| Magnesium | Bismuth | Copper | Iron | Copper |
| Zinc | Aluminium | Tin | Nickel | Lead |
| Aluminium | Structural steels | Lead | Copper | Tin |
| Lead | Zinc | Zinc | Aluminium | Nickel |
| Tin | | Iron | Tungsten | Iron |
| Bismuth | | | Zinc | Zinc |
| | | | Tin | |
| | | | Lead | |

**Toughness** is the property that enables a material to withstand shock and to be deformed without breaking.

**Brittleness** is a property of a material that makes it break easily without bending. Cast iron and glass are good examples of brittle materials.

**Malleability**. A malleable metal can easily be deformed, especially by hammering or rolling, without cracking.

**Ductility** is the property that enables a material to stretch, bend or twist without cracking or breaking. Ductility is the opposite of brittleness.

**Corrosion resistance** is the ability of a metal to withstand **corrosion**.

**corrosion** – deterioration of a material from exposure to a particular environment or subjected to a particular type of behaviour

The mechanical properties tell an engineer many of the things he or she needs to know when considering which material to use for the body of a car.

## Spot welding

However, it is not just the section of material that affects strength. In many cars, the cross sections are spot-welded together. Spot welding is a common method, particularly in the motor manufacturing industry, of joining thin sheets of metal together. A spot weld is achieved by pressing together two thin-walled sheets of steel between copper electrodes. By passing a large electric current between electrodes for a short period of time, heat is generated. The electrical resistance of the two sheets causes some of the steel to melt and join. The name given to the join is a nugget.

To obtain a good quality nugget, the sheets need to be clean and free from grit, dirt, water and oil. It is also necessary to supply the right amount of electric current, for the correct length of time. Applying the current for too long can burn a hole right through the materials being welded.

It should be noted that using spot welding results in a join occurring only where the spot weld is. In rally cars these joins are often strengthened by adding a continuous weld.

While it would be desirable to have the spot welds very close together to gain extra strength, this is not possible as, if the spot welds are too close, the electric current flows through an existing spot weld rather than creating a new one. Engineers refer to this as shunting, and the effect is to reduce the size of the second nugget and thus reduce the strength of the joint. Most spot welds are created using robots. Although there is a tendency to describe robots as keen tireless workers, robotic systems can have bad days too. With quality control dependent on factors such as the force of the electrodes, the shape of the electrodes and the amount of current, it is easy to see how variables can affect the efficiency of the robot. It is common to see on TV spot-welding robots working hard with sparks flying everywhere. While this may look spectacular, sparks are actually the sign of a bad spot weld as they are caused by small particles of hot metal being fired out of the nugget – weld splash. Any loss of metal will result in a reduced spot weld and thus reduced strength.

**velocity** – the speed at which an object is travelling

In terms of Jaguar, the Jaguar XK Coupé body has only one single welded joint – in one cosmetic joint in the roof. This reduction in the number of welded joints increases the strength of the car – an important safety consideration.

## The science of a car crash

**MATHS & SCIENCE**
Newton's First Law of Motion.

An object at rest tends to stay at rest, and an object in motion tends to stay in motion with the same speed and in the same direction unless acted upon by an unbalanced force.

**MATHS & SCIENCE**
Newton's Second Law of Motion states that an object that is acted on by a constant force will move with constant acceleration in the direction of the force.

In order to accelerate (or decelerate) a mass, a force must be applied. The way it is often expressed is with the equation:

$$F = ma$$

The force 'F' is what is needed to move a mass 'm' with an acceleration 'a'. Acceleration has a defined mathematical definition relating to the increase in **velocity**. Acceleration has nothing to do with going fast. A fast car can be moving very fast, and still not be accelerating. If an object is not increasing its velocity, then the object is not accelerating.

**MATHS & SCIENCE**
To convert mph to km/h multiply by 1.6093. To convert km/h to mph divide by 1.6093.

If a car is travelling at 48 km/h (30 mph), you are also travelling at 48 km/h inside the car so you feel as if you and the car are moving together.

If the car were to crash into a stationary object, the force of the object would bring the car to a sudden stop; however, your speed would remain the same. You would naturally continue in a forward motion unless you were acted upon by the unbalanced force of a seat belt.

Just as the stationary object slowed the car down, your seat belt would slow you down by exerting a tremendous amount of force.

**MATHS & SCIENCE**
Inertia is the property of an object to keep moving unless acted upon by an outside force. It is defined using Newton's First Law of Motion.

You are thrown forward in a crash because of the principle of **inertia** – the property of an object to resist any change in its state of motion.

If you are not wearing a seat belt, the dashboard, windscreen or road would slow you down by exerting a force. However, impact with these objects can severely injure or kill a person.

### What happens in a crash?

When a vehicle crashes, a large force, or energy, is generated. Energy

FIG 2.4: What can happen if you don't wear a seat belt?

cannot be cancelled: it must go somewhere. This is known as the **law of conservation of energy**: energy cannot be created or destroyed; it may be **transformed** from one form to another, but the total amount of energy never changes.

**transformed** – changed completely

In a car crash, the energy, known as crash energy, does the **work** that crushes the car's crumple zones.

MATHS & SCIENCE

Work is the ability to apply a force (push or pull) over a distance. W = F x D. The work done (W) is equal to the force (F) multiplied by the distance (D) the object travels.

Let us now imagine two cars heading towards each other.
Assume that the cars are identical, are of the same weight and travelling at the same speed. If they collide head-on, the force created will be double that experienced by, for example, a single car hitting a rigid wall and, importantly, the force will be distributed evenly on to each car. So the effect of the force would be the same for each car, and it would be as if each car had crashed against a rigid wall.

If, however, one car was much bigger than the other, the effect of the force would be much worse for the smaller car. This is because the proportion of the force to the car's mass would be much greater.

**momentum** – the force that keeps an object moving

The **momentum** of a moving object is related to its mass and velocity. A moving object has a large momentum if it has a large mass, a large velocity or both. A smaller car can be stopped more easily than a larger car. Both cars have momentum but the larger car has more momentum.

MATHS & SCIENCE

The momentum of a body is its mass multiplied by its velocity.

Momentum changes if the velocity and/or mass changes, so increasing an object's mass increases its momentum or 'bashing power'.

## Anatomy of a pedestrian crash

Most pedestrian crashes involve a forward-moving car. In such a crash, a standing or walking pedestrian is struck and then continues forward as the car brakes to a halt. Although the pedestrian is impacted twice, first by the car (primary collision) and then by the ground (secondary collision), most of the fatal injuries occur due to the interaction with the car.

The diagram below shows the results of a series of tests to replicate accidents involving children and adult pedestrians at 40 km/h (25 mph).

**FIG 2.5:** Contact areas as a result of a car hitting a pedestrian

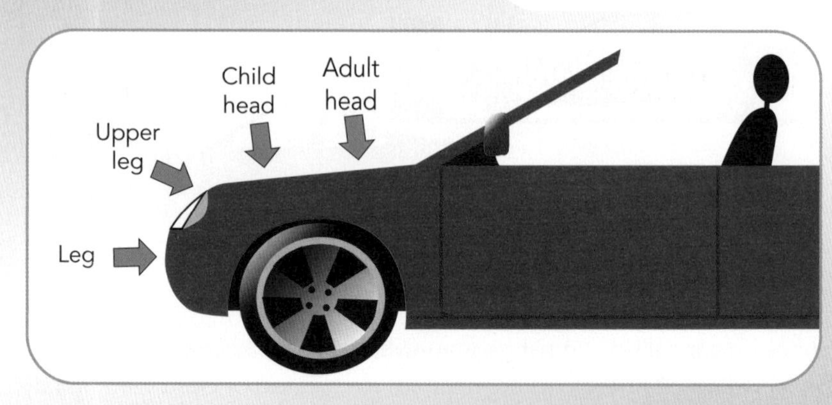


**stature** – a person's natural height

The location of the head impact is dependent on pedestrian **stature** and motion, the position of impact across the width of the car, and the size and shape of the vehicle involved. A large area of the bonnet top can potentially be hit in pedestrian accidents.

Head injuries caused in primary collisions account for about 80 per cent of pedestrian deaths.

Thus, vehicle designers usually focus their attention on understanding the car-pedestrian interaction, which is characterised by the following sequence of events:

Pedestrian dummy testing pop-up bonnet system

Firstly, the front bumper strikes the pedestrian's leg in the calf and knee area. Then the pedestrian's thigh or pelvis is struck by the leading edge (front edge) of the bonnet. After this the pedestrian rotates about the leading edge until their arms, head and chest strike the bonnet top and/or windscreen.

On the Jaguar XK the design of a passive bumper system helps **mitigate** leg injury through the use of crushable foam and plastic covering.

**mitigate** – make something less severe, serious or painful

**aerodynamics** – the study of the interaction between air and the solid bodies moving through it

## Protecting the head

The bonnet of most vehicles is usually made from sheet metal (as previously discussed), which is an energy-absorbing structure and thus poses a comparatively small threat. Most serious head injuries occur when there is insufficient clearance between the bonnet and the stiff underlying engine components.

Creating room under the bonnet is not always easy because there are other design constraints such as **aerodynamics** and styling.

The pop-up bonnet works by rising up on impact with a pedestrian, thereby cushioning the pedestrian from the hard engine parts. The pop-up bonnet is designed to create more space between pedestrians and the engine and thus reduce the severity of injuries to pedestrians in the event of a collision with a car.

**FIG 2.6:** Reduction in severity of head injuries as a result of the deployable bonnet system

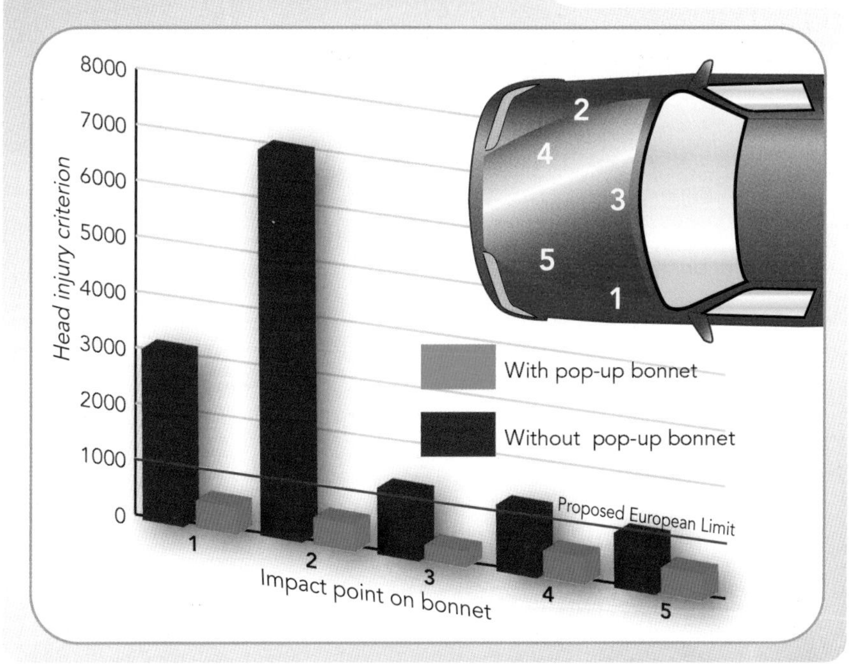


Sensors along the front part of the bumper can differentiate a pedestrian from other objects, e.g. birds. This is done by considering the mass and

stiffness of the object. Once a pedestrian is detected, a signal is sent to the electronic control unit (ECU), which opens the latches and fires up two miniature air bags. The bonnet is raised up by 5 inches (about 130 mm), creating an air gap between the under-bonnet components and the pedestrian, as well as giving a cushioning effect.

Jaguar's pop-up bonnet system can lift the bonnet (which weighs 18 kg) in around 30 milliseconds, faster than you can blink, which requires an acceleration rate of about 50 times the force of gravity. A small **pyrotechnic** charge raises the bonnet using a raising mechanism. In a typical collision at 40 km/h (25 mph), the victim's head hits the bonnet about 150 milliseconds after the bumper first hits their legs, by which time the pop-up bonnet has been activated, helping to decelerate the pedestrian's head and thus lessen levels of injury.

**pyrotechnic** – relating to fireworks

As a result of this system Jaguar won the 2006 World Traffic Safety Symposiums Traffic Safety Achievement Award.

**NUTS & BOLTS**

**To find out more about measures relating to the protection of pedestrians go to http://ec.europa.eu/enterprise/automotive/pagesbackground/pedestrianprotection/final_trl_2006.pdf**

Air bags lift up the bonnet

Bonnet-raising mechanism developed by Jaguar

# The process

## Planning

### The brief

Since the mid 1990s European legislation has been trying to tackle the problem of pedestrian safety. In the same way that a car must pass European safety standards for occupant protection in front and side impacts, cars sold in Europe are also subject to a pedestrian impact test. While no vehicle to date has been withdrawn from sale for a poor pedestrian score (unlike a poor crash test score), the negative image associated with such has caused car manufactures to progressively change the design of their cars to fare better in **simulated** tests.

However, the need to comply with Phase One of new European safety legislation (a set of test procedures for evaluating the safety performance of cars when they strike pedestrians) and, in particular, required improvements to the bonnet area to make it safer for head impacts have provided the motor manufacturing industry with a problem.

**simulated** – to do or make something which looks real but is not real

**patent** – the official legal right to make or sell an invention for a particular number of years

### The design

When developing new ideas, in general, there are three areas that Jaguar will consider:

- legislation – are there specific requirements that need to be complied with?
- competition – have other motor manufacturers developed a new component?
- innovation – can a unique component be developed that delights the potential customer?

In this particular instance, the need for design is as a result of new safety legislation, which has a fixed time frame for compliance and successful implementation.

Jaguar must check whether its idea is new and original, i.e. no one else has a similar product on the market. If it is an original idea, Jaguar would need to apply for a **patent** which would grant them sole rights to that idea.

Once they have developed their design ideas they must consult with their supply base to consider the different alternatives that may be available to them. It is important to consider how the new component will interact with everything else on the car. The Jaguar design team will then provide a specification for the new component which must meet the standard Jaguar criteria of heat, humidity and durability.

After a product specification is finalised, the component is then produced as a **virtual build**, i.e. a computer simulation. Extensive computer analysis will be undertaken to check that the component works and operates in the correct way, i.e. does it do what it was designed to do?

**virtual build** – not physically existing as such but made by software to appear to do so
**prototype** – first or preliminary version of a device from which other forms are developed
**'real-world'** – actual reality

## Production and testing

Initially the new component is manufactured as a **prototype**. This prototype is then tested as a component part, i.e. it is tested as a stand-alone product, and then it is tested as part of the car.

Once the prototype has satisfactorily passed all of Jaguar's testing procedures it will enter the full production process and actual production line build.

In real terms, the complex pop-up bonnet system has been extensively researched across wide-ranging scenarios, using 120 man-years and thousands of computer simulations, as well as having been tested in practice at Jaguar's Engineering Centre at Whitley in Coventry, England. While all pedestrian impact research has been carried out using virtual tools, analysis of previous '**real-world**' incidents has played an important part in the development process.

Jaguar's Programmes Director said: 'I take technology development at our product development centres very seriously, and I am proud to be working with Jaguar engineers who can deliver this kind of safety system, which we believe to be a world first'.

### Car test procedures explained

To provide a full picture of car safety we will now consider the tests that the cars must undergo.

You will be aware of the crash test dummies that are used for the test simulations. Each of the dummies used mimics the human shape, size and weight and is tailored for a specific test. Adult frontal-crash dummies consist of a small female, mid-size male and a large male. Child dummies include 6-, 12- and 18-month-old babies, and 3-year-old, 6-year-old and 10-year-old siblings. A different set of dummies are used for side impact tests, and these are known as the SIDs family (side impact dummies).

The data collected from the dummies following crash testing not only provides information on impact injuries to specific body parts, but is also used to support computer modelling, allowing the construction of precise virtual models. These virtual models can then be used in computer crash simulations.

**FIG 2.7:** Frontal offset crash test

Frontal offset tests are based on those developed by the European Enhanced Vehicle-safety Committee as a basis for legislation.

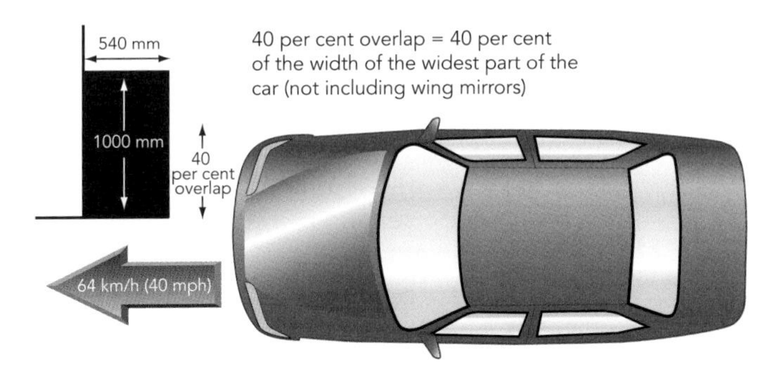


The vehicle is run into the barrier at 64 km/h (40 mph) in order to measure the impact on the head, chest and legs of the dummy, as well as to check the condition of the deformed car.

This test represents the forces involved in a typical head-on collision of two vehicles weighing the same and travelling at 64 km/h (40 mph).

Readings taken from the dummies are used to assess the protection given to adult front occupants.

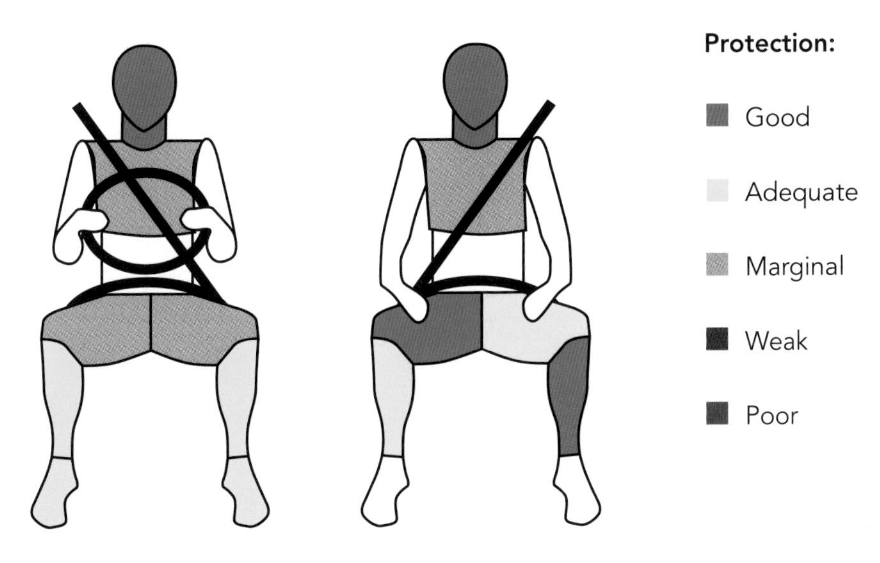

**Protection:**

- Good
- Adequate
- Marginal
- Weak
- Poor

Accident patterns vary from country to country within Europe, but approximately a quarter of all serious-to-fatal injuries happen in side impact collisions. Although this is much less than frontal impacts, injuries per accident are much higher. Many of these injuries occur when one car runs into the side of another. The main reason for increased injuries here is that in a front-end accident the car occupants are protected by the car engine, the front wheels and the engine mounting frame. The crush length of these can be around one metre compared to about 2 cm (the thickness of the car door) when it comes to side impacts.

**FIG 2.8:** Side impact tests

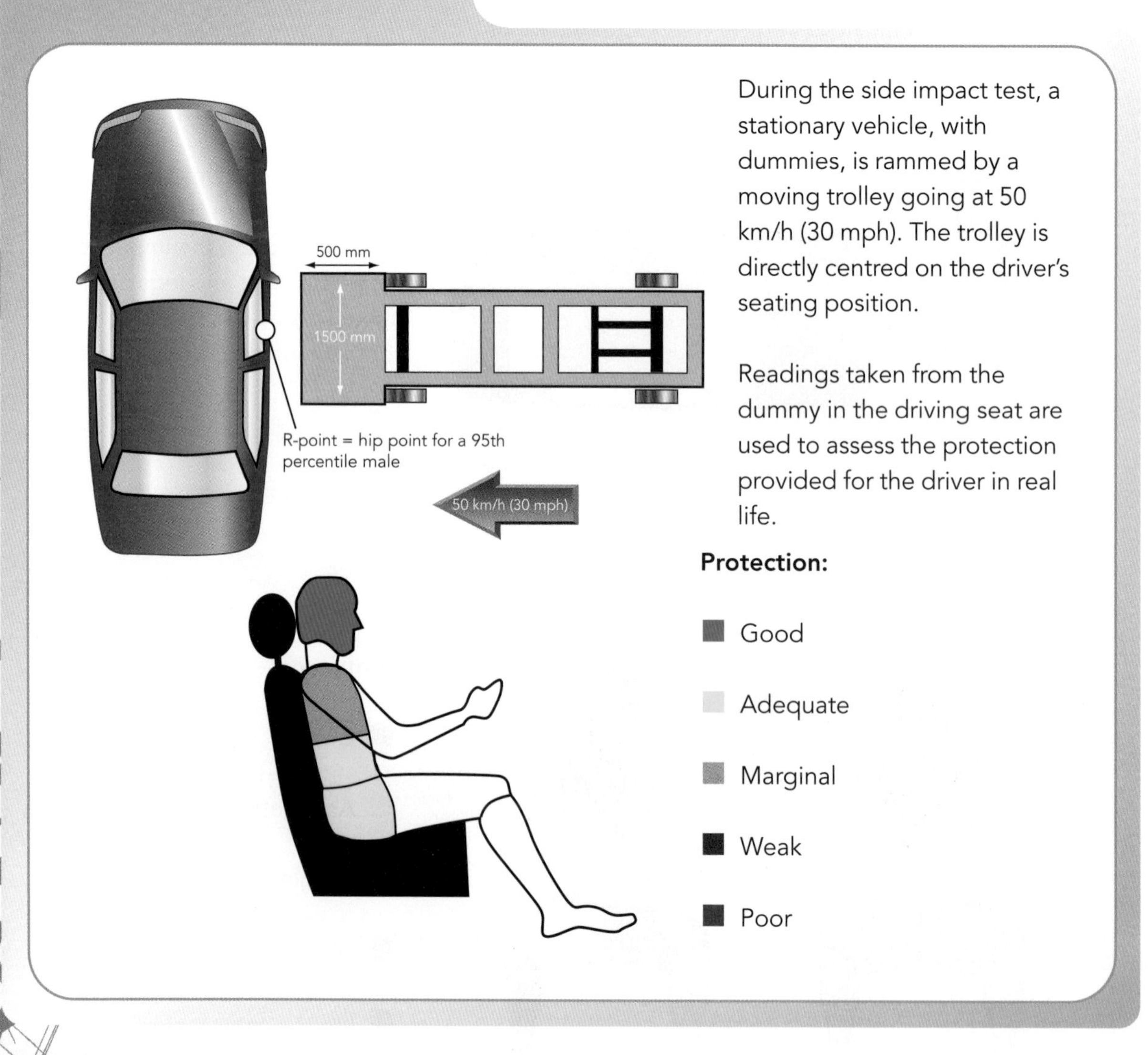

# The career path

## PERSONAL PROFILE: Bill McLundie

**Qualifications**
PhD – part-time, due to be completed in 2007.
MSc in Aerospace Vehicle Design.
BSc (Eng) in Aeronautical Engineering.

**Professional affiliation**
The Royal Aeronautical Society, MRAeS (Chartered Engineer)

**Professional experience**
*Principal Technical Specialist* *Premier Automotive Group (Jaguar/Land-Rover)*

- Created systems specification and led development and testing of the world's first active deployable bonnet system.
- Involvement in the development of Ford's global safety strategy.
- Negotiating at a European industry level on future test procedures.

*Team Leader* *Premier Automotive Group (Jaguar/Land-Rover)*

- Responsible for overseeing all new advanced engineering projects on future vehicle programmes for body and trim activities.

*Project Engineer* *Pedestrian Research* *Jaguar Cars*

- Internal consultant on vehicle design.
- Involvement in the Ford 'Big Bang' research project – first use of an accurate LS-DYNA pedestrian humanoid on a Ford vehicle programme.

*Senior Engineer* *1999–2000* *Jaguar Cars*

- Investigating structural safety, particularly rear crash implications for XJ.

*Team Leader* *1996–1999* *BAe Airbus Ltd*

- Development of IT and business processes and methodologies for aircraft programmes.
- Led the initial support team on the A380.

### What made you want to be an engineer?

The fascination of how things work. I was one of those children who took things apart and tried to put them back together again.

### What are the benefits of working for Jaguar?

Being able to drive the new cars 12 months before the journalists, such as Jeremy Clarkson.

**To find out more about careers in the motor industry go to www.automotive-skills.org.uk/auto/control/Careers and www.learndirect-advice.co.uk/helpwithyourcareer/**

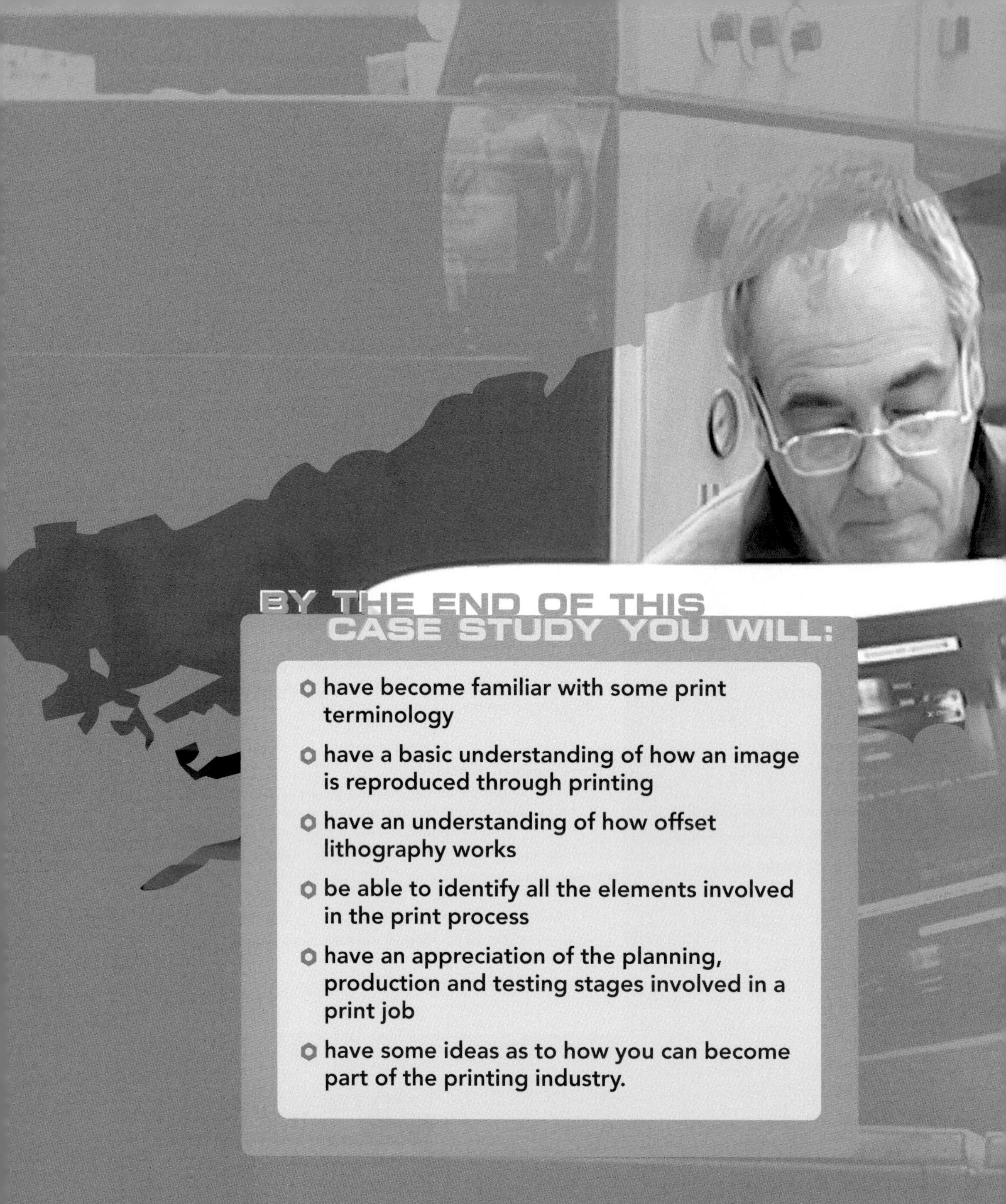


## BY THE END OF THIS CASE STUDY YOU WILL:

- have become familiar with some print terminology
- have a basic understanding of how an image is reproduced through printing
- have an understanding of how offset lithography works
- be able to identify all the elements involved in the print process
- have an appreciation of the planning, production and testing stages involved in a print job
- have some ideas as to how you can become part of the printing industry.

# THE PRINT INDUSTRY: WYNDEHAM WESTWAY

## Case study overview

This case study introduces you to the printing industry and the individual elements that contribute to the final print product.

Over the following pages you will become familiar with a real company that has the capability of printing up to one million sheets per day or, in actual terms, 16,000,000 double-sided A4 sheets.

After considering how the actual print processes work, you will follow the company's procedures from initial client brief through to the final print product. This will help you to understand and appreciate all the steps involved in achieving the finished product.

Finally, you will be introduced to a company employee who will describe the route she took to her current position.

# The sector

## The printing industry

Printing is part of the papermaking and paper products sector. This sector covers all aspects of the production of paper, board and tissue products that you come into contact with each day: the morning newspaper, adverts, books, magazines, flyers and directories, even your breakfast cereal packets and the exam papers that you sit at the end of your course.

**multinational** – operating in several countries

The printing industry is made up of more than 12,000 companies, which range in size from major **multinational** companies to small family-run businesses employing less than five workers – approximately 90 per cent of companies in the printing industry employ less than 50 people.

The printing industry is a major contributor to the economy, with the value of sales being worth over £14 billion. It is ranked fifth largest within the UK economy.

The industry serves all sectors of the economy, including public authorities, financial services, publishers, distributive services and the manufacturing industry. Everyone handles its products, often without giving their producers or processes a second thought.

> The printer must put his labour and materials to the best possible use in disseminating knowledge more widely, conveying news more speedily, expressing information more clearly, commemorating events more gracefully, and adorning life more beautifully.
>
> Winston Churchill

## The products

The printing industry is not solely concerned with the production of newspapers and magazines. Products cover everything from greetings cards to envelopes through to printed products for supermarkets and every kind of packaging, including labels, wrapping, cartons and containers.

Other print products include currency, cheques, credit cards and other security products, such as utility and credit card bills. Printing is carried out on every kind of material and surface – plastic, film, beverage containers, metal and wallcoverings.

# The company

## Background

The Wyndeham Press Group was formed in 1991 and has developed through **acquisitions** and **organic growth** into one of the UK's leading printing groups. Services offered include printing, finishing and mailing, design, and **pre-press** and packaging.

**acquisitions** – buying or obtaining of assets or objects
**organic growth** – gradual or natural development
**pre-press** – composition, page layout and other work done on a publication before it is actually printed

The Wyndeham Press Group has an annual turnover of £140 million and 12 sites across the UK.

Wyndeham Westway, based in Luton and employing 92 staff, is one of the UK's largest commercial **sheet-fed** print facilities. It has an annual turnover of £10 million.

**sheet-fed** – a press which prints by taking up one sheet at a time

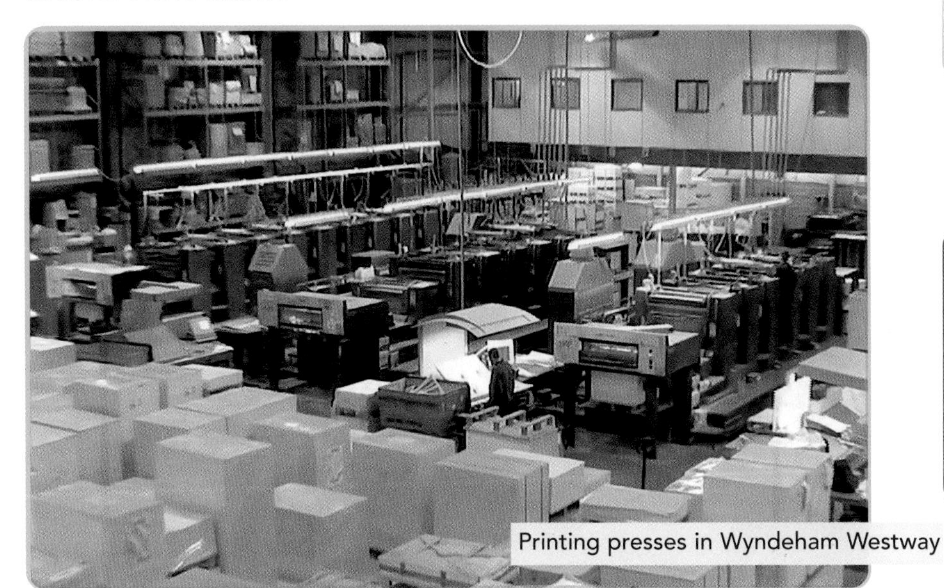
Printing presses in Wyndeham Westway

**NUTS & BOLTS**

To find out more about the Wyndeham Press Group go to www.wyndeham.co.uk

**B1** – paper size 707 x 1000 mm

**financial viability** – having enough revenue coming in and cash on hand to pay bills or debt

Westway has five **B1** presses at an investment of £1.5–2 million per press. Constant upgrading and maintenance needs mean that the B1 presses must be working at full capacity for 24 hours a day, almost seven days a week. This is to ensure **financial viability** in extremely difficult market conditions, where Westway has less than one per cent of the market share but is still one of the top ten B1 commercial printers.

Westway is capable of printing a diverse range of marketing materials – from business cards, brochures, leaflets and books to catalogues and magazines. It can print from two up to ten colours on a range of paper sizes.

This modern, purpose-built plant has the capability of printing up to one million sheets per day, which, in actual terms, represents 16,000,000 double-sided A4 sheets!

HEALTH & SAFETY

**High energy use within the print industry primarily originates from its dependence on large, heavy machinery. The increase in the amount of energy used globally has been linked with the increase in the Earth's surface temperature.**

## Modern-day concerns

The print industry is the fifth largest and the sixth worst polluting manufacturing industry in the UK. By its nature, printing is an extremely **resource-intensive** activity, consuming large quantities of energy and forestry.

ISO 14001 is an internationally accepted standard that sets out how a company can go about putting in place an effective Environmental Management System. The standard is designed to address the delicate balance between maintaining profitability and reducing environmental impact.

**resource-intensive** – needs a lot of materials to operate

**accreditation** – official recognition, acceptance or approval of something when recognised standards have been met

The Wyndeham Group is committed to its role as an environmentally conscious manufacturer and, accordingly, it has put in place an environmental management team to enable ISO 14001 **accreditation** across every site within the Wyndeham Press Group. Wyndeham Gait was the first company within the group to be awarded ISO 14001 in early 2004.

# The product

## Introduction

Print production is now almost a totally digital process and, as such, has a high reliance on computer equipment. Data storage and connectivity between different devices is an essential part of the printing process.

## How does printing work?

If you draw a picture using a single crayon or a coloured pencil, you can produce a **continuous tone** image by colouring in your drawing using different pressures on the pencil, or by adding extra colour over the top of existing colour. The name given to your image is contone (continuous tone).

**continuous tone** – image made up of smoothly graduated tones or colours or shades, rather than solid blocks of colour

The change from one shade of a colour to another is gradual and smooth. In your image the colour will always, basically, be the same, it will just be an image of varying shades of the same colour. If you apply more colours, you will achieve the same effect, only in a more complex way.

Printing presses typically cannot reproduce continuous tone pictures directly since they use only one or four or, occasionally, a handful of colours of ink (e.g. spot colour). Continuous tone pictures are thus simulated through the use of halftones.

**Halftone** is a method of producing a range of tones by dividing the image into a series of dots. A halftone screen – a transparent material comprised of a finely ruled grid of lines – is used to translate the full tone of the image into a series of dots.

FIG 3.1: Halftone image

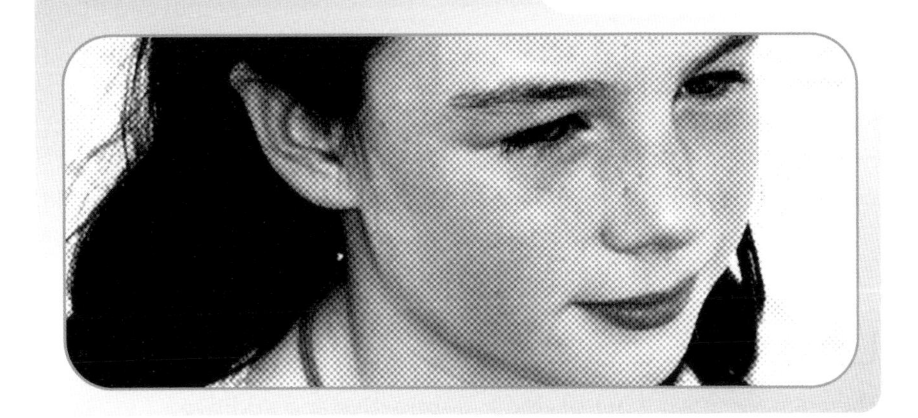

**MATHS & SCIENCE**

**A mixture of imperial (inch) and metric (millimetre) measures are used within the print industry.**

**coarse** – of low or inferior quality

**pixel** – the smallest unit of an image on a television or computer screen

**resolution** – the degree of detail visible in an image

Dark areas have relatively big dots, close together. Light areas have small dots surrounded by white space. The number of dots used determines the quality of the image produced. In a newspaper, the halftone dots are easily visible to the naked eye – the screen used can often be as **coarse** as 60 **dpi (dots per inch)**. A colour magazine would typically use a screen of 150 dpi, an art book possibly 175 dpi or finer.

A halftone screen can also be applied to a solid colour in order to produce tints of that colour. In this instance, the screen contains all the same size dots.

Using the same concept, an image on a computer or television screen is made up of a series of dots known as **pixels**. The smaller the pixel, the more detailed the picture or the greater the **resolution**.

If the number of pixels is too low, you can see the dots. The name given to this is **pixelation**. High-resolution images have a much larger file size.

FIG 3.2: High- and low-resolution images

If you look at the two images above, you will notice the difference between the two resolutions. In printing, you will commonly find images produced in both low and high resolution. The low-resolution image will be used to communicate with a client over the internet, as the smaller file size will allow a much faster transfer of the document. The high-resolution image will be used for the actual printing process.

## Use of colour

There are two main types of colour: **subtractive** and **additive**.

## Subtractive primary colours

Subtractive primary colours are used when we are dealing with **reflected light**. Because of this we use them when we are mixing paints or inks. The subtractive primary colours are cyan (a light blue), magenta (a pinkish purple) and yellow.

You will probably be thinking that in art, at school, you were taught that the subtractive primary colours are red, blue and yellow. This is incorrect. So how has the confusion arisen?

At school, when you want to create a colour that you do not have, you are told that you should mix the right proportions of each of the subtractive primary colours. It is very rare, however, that you will find cyan or magenta paint or crayons amongst the selection given to you, so your teacher helpfully advises you to use red as an approximation to magenta, and blue as an approximation to cyan. From that moment on you will believe that you can mix red, yellow, and blue to make any colour, and that they must, therefore, be primary colours.

The vast majority of magazines and colour books are produced using a **four-colour process**. The four ink colours used by the printers are cyan, magenta, yellow and black, referred to as **CMYK**. Because the inks used are **translucent**, they can be **overprinted** and combined in a variety of different proportions to produce a wide range of colours.

FIG 3.3: Subtractive primary colours

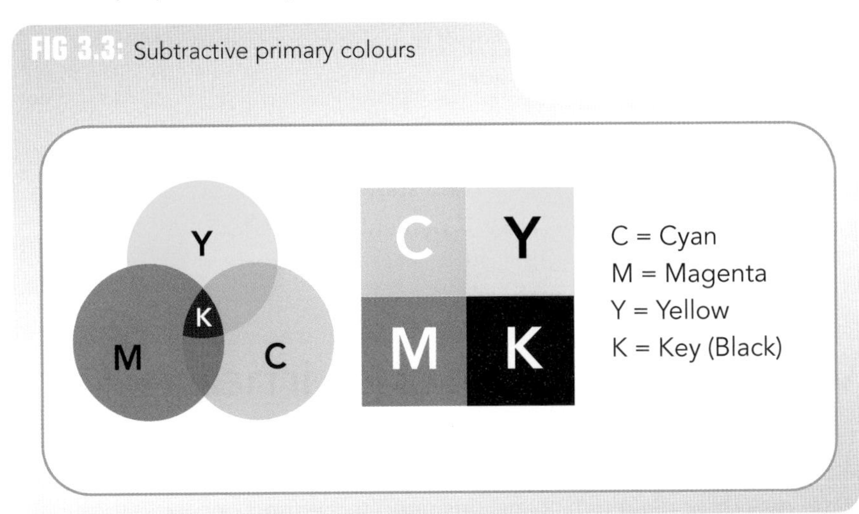


These are called subtractive colours, as combining them all gives the colour black. Subtracting one or more of these colours will yield any other colour. When combined, in various percentages, these four inks will create an entire **spectrum** of colours.

Theoretically, it is possible to produce an adequate range of colours using just cyan, magenta and yellow. However, in practice much better results are achieved with the addition of black. The black plate is used to strengthen the shadow areas and reduce the amount of CMY inks required.

**MATHS & SCIENCE**

**Visible light is made up of 7 wavelength groups. When light hits objects, some of the wavelengths are absorbed and some are reflected. The reflected wavelengths are what we 'see' as the object's colour.**

**translucent** – semi-transparent
**overprinted** – print additional matter on a surface already bearing print
**spectrum** – a band of colours, as seen in a rainbow

## Additive primary colours

Additive primary colours are used when we are dealing with mixing **emitted light**. The additive primary colours are red, green and blue.

All colours can be produced by mixing these three colours. An example of their use is on a computer monitor, where varying intensities of red, green and blue light are used to create the colours we see. If full-intensity red, green, and blue are mixed we get **white light**.

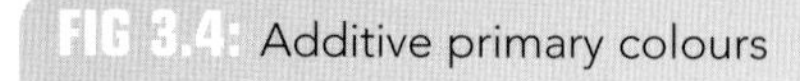

FIG 3.4: Additive primary colours

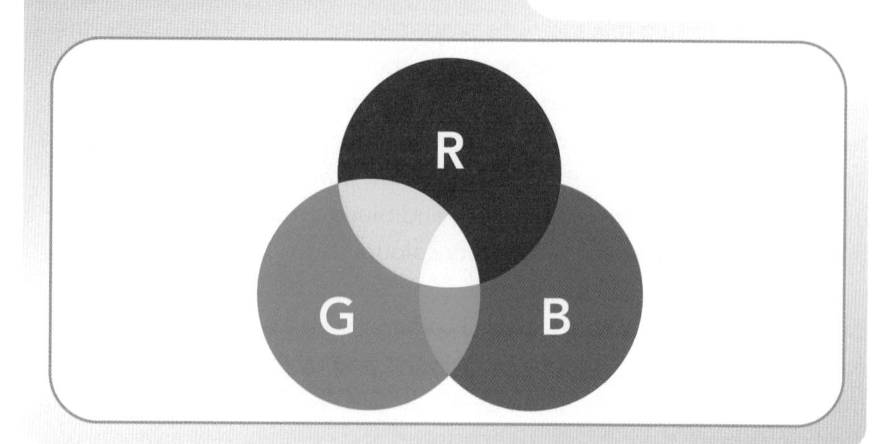


**MATHS & SCIENCE**

In 1666, Sir Isaac Newton studied the effects of light passing through a prism. The light separated into a rainbow of colours, which he passed through a second prism to form white light.

## Colours and the printing industry

Any area of colour that is not printed using the four colour (CMYK) process is known as a **spot colour**.

Spot colour is the name given to a custom-mixed ink colour used in printing. Pantone is a commonly used spot colour matching system. The Pantone system allows users to mix percentages of base inks (such as CMYK) to create new colours.

# What carries the image?

The image is carried by a plate known as the **printing plate**. These can be made from a variety of materials. At the cheaper end of the market there are paper plates which are designed to be used once and thrown away. These are very economic for short runs, such as small quantities of stationery.

## Offset lithography

Offset lithography is the most common method of commercial printing at this time.

Artwork is produced digitally with graphic design software. Traditionally, a film (negative) was produced first and the plate was made from the film.

FIG 3.5: Artwork to plate process

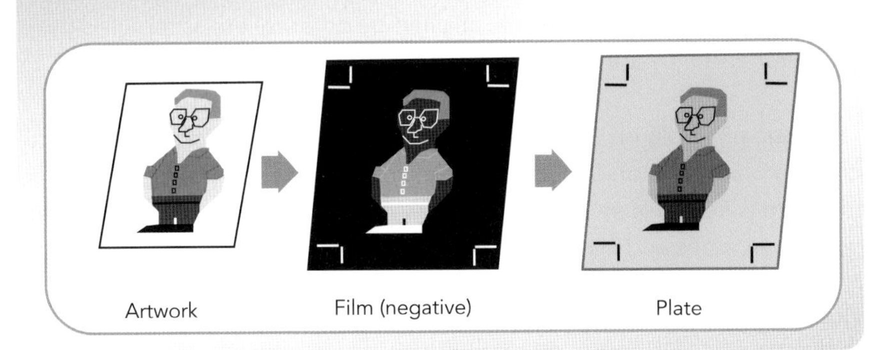


Images from the negatives are transferred to the printing plates in much the same way as photographs are developed. A measured amount of light is allowed to pass through the film negatives to expose the printing plate.

When the plates are exposed to light, a chemical reaction occurs that allows an ink-receptive coating (**photosensitive emulsion**) on the plate to be activated. This results in the transfer of the image from the negative to the plate.

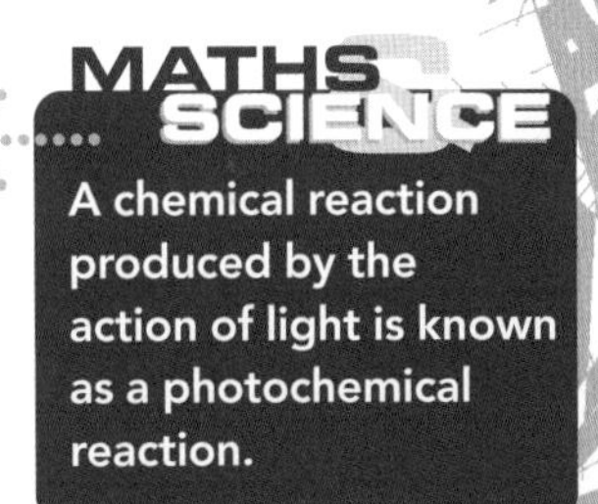

**MATHS & SCIENCE**

**A chemical reaction produced by the action of light is known as a photochemical reaction.**

Today, it is possible to do this without a final film, using a system called computer-to-plate (CTP).

When printing with more than one colour there is one separate plate for each ink used.

FIG 3.6: Offset lithography

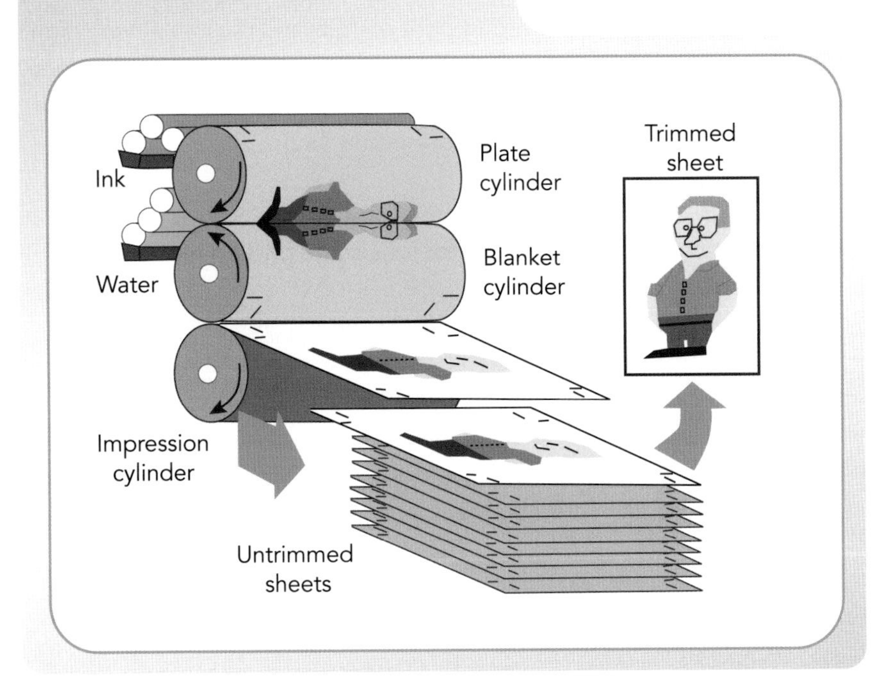

The plate is affixed to a drum on a printing press. Rollers apply water, which covers the non-image areas of the plate but is repelled by the emulsion on the image area. Ink, applied by other rollers, is repelled by the water and only adheres to the emulsion of the image area.

The basic principle on which it works is that oil and water do not mix. The printing plate has non-image areas which absorb water. During printing the plate is kept wet so that the ink, which is inherently greasy, is rejected by the wet areas and adheres to the image areas.

The image is not directly transferred to the paper as the paper would become too wet. Instead the plate rolls against a rubber blanket, which squeezes away the water and picks up the inks. The paper rolls across the blanket drum and the image is transferred to paper. This indirect method is the **offset**, after which the process is named.

**finishing** – any process that follows the actual printing

After printing the sheets are taken for **finishing** – trimming, folding and binding.

# Other printing techniques

## Flexography printing

Flexography printing is a form of **relief printing** – where the parts to be printed are raised up from the base plate.

Flexography uses a relief image and thin, flexible printing plates made of rubber or photopolymer. The image on the plates is produced by a photographic process, and the inks are quick-drying thin liquids.

This printing method is still used to print on foils and plastics at very high speed. Flexography can be used to print on materials such as cellophane, polythene and metallic films, so the technique is used for printing plastic shopping bags and packaging for food products.

## Gravure process

Gravure printing is commonly used for labels and packaging.

For this process, the image to be printed is made up of small holes sunk into the surface of the printing plate. The entire surface of the plate is inked and the excess is scraped off with a knife, which leaves the ink in the depressions (the small holes) of the plate. A rubber-covered roller presses paper onto the surface of the plate and into contact with the ink in the depressions, thus printing the image.

The printing plates are usually made from copper and may be produced by engraving or etching.

## Screen printing

Screen printing is very similar to stencilling.

This is a process where the ink is transferred to the printing surface by being squeezed through a fine fabric sheet stretched on a frame. The screen carries a stencil which defines the image area.

There are several ways to create a stencil for screen printing. The simplest method is to create it by hand in the desired shape, either by cutting a piece of paper or plastic film and attaching it to the screen. An increasingly popular method is to create a stencil photographically. Photographic screens can reproduce images with a high level of detail, and can be reused for thousand of copies.

Compared to other types of printing, screen printing is slow and very labour-intensive. But it is possible to screen-print an image onto a wide range of textiles and other materials. It is ideal for printing posters, T-shirts, display boards, fabric and wallpaper.

# How is the printed paper folded?

Once printed, leaflets and many other print products need to be folded accurately.

Folding involves the use of mechanical pressure to produce a high-quality sharply defined fold.

In mechanical folding, the paper or card is doubled up between two rollers while pressure, appropriate to the thickness of the paper, is applied. The end result is a sharp fold that virtually eliminates the paper's natural tendency to return to its original shape.

There are two main types of folding machine: **buckle folders** and **knife folders**. One type of folding may be used alone, or both techniques can be combined.

In a buckle folder, the front edge of a moving sheet of paper or card is stopped allowing it to **buckle** at a set point. The system needs three rollers and a buckle plate. The first two rollers are arranged vertically above one another and their job is to carry the incoming paper or card towards the buckle plate until it reaches an adjustable stop. As the lead edge of the sheet strikes the feed guide stop, the sheet continues to be fed into the buckle plate, creating a buckle in the space between the rollers. As the excess paper is pushed downwards it is grabbed by the **contra-rotating** rollers, and the fold is formed as the sheet passes through them.

**buckle** – bend and give way under pressure or strain

**contra-rotating** – rotating in opposite directions

FIG 3.7: Buckle folding of printed paper

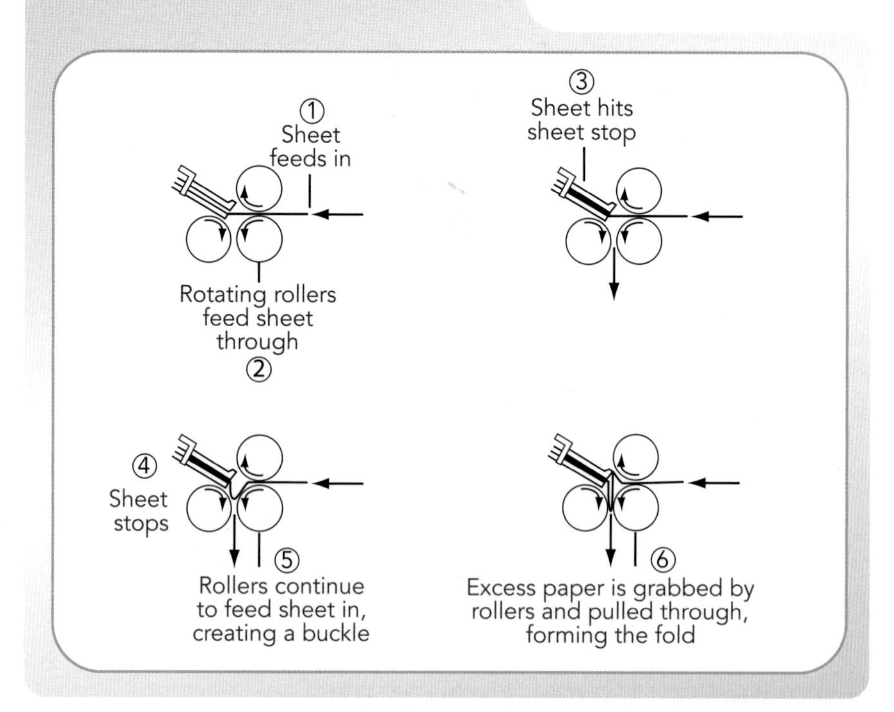


Knife folding uses a knife to force the paper, at a specific point, between two rollers turning in opposite directions. The sheet is carried to the folding station until it makes contact with the sheet stop. The knife then descends vertically, pushing the sheet between the two rollers that have been set to the thickness of the sheet going through them. As the sheet passes through the rollers, it is pinched and the fold is formed.

The knife folder is a slower-speed folder, but it has greater accuracy when producing right-angle folds.

FIG 3.8: Knife folding of printed paper

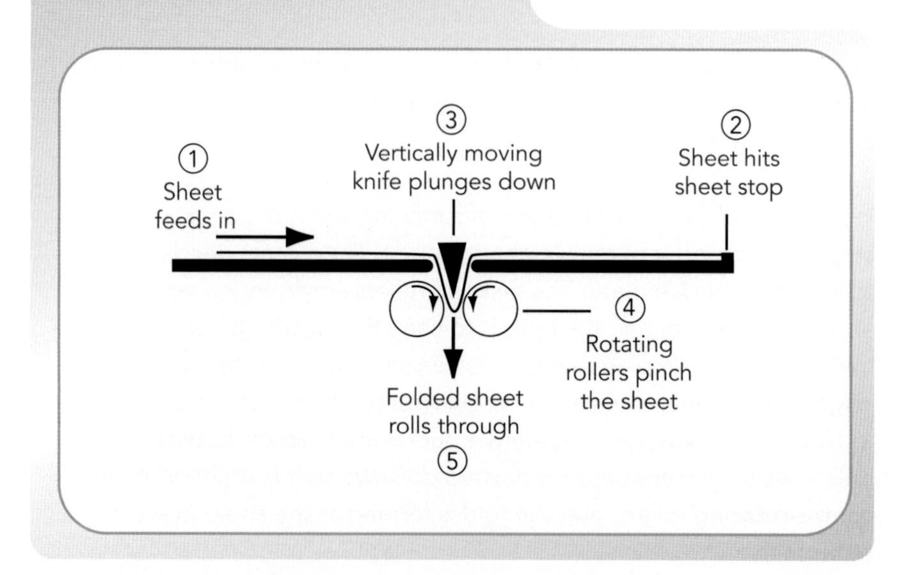

# Critical points

When a number of folded pages are fitted together and stapled (saddle stitching) the printer has to ensure there is no **creep**.

The bulk of the paper causes the inner pages to extend or creep further out than the outer pages when folded. You may think that this is not a problem, as all you need to do is trim the overhanging edges. However, the printed image on the inner pages will also be moving further and further towards the edge.

Creep varies depending on the thickness of the paper and the number of pages. If there is no creep **allowance**, when pages are trimmed, the outer margin becomes narrower towards the centre of the booklet and there is the possibility that text or images may be cut off.

Another issue for printers is print bleed. **Bleed** is the printed area which extends off the trimmed area.

It is not possible to print all the way to the edge of the sheet of paper. To achieve this effect it is necessary to print a larger area than is required and then trim the paper down. Typically, an extra 3 mm of bleed is allowed for colour and image areas, to allow for a tolerance in the accuracy of cutting.

In terms of folding, the direction of the paper grain must be considered. Paper that is folded **parallel** to the grain of the paper will fold much more cleanly. A cleaner fold is produced because the paper fibres (grain) are running in the same basic direction as the fold. Only a few of the fibres provide any resistance to the folding action, which results in a high-quality fold.

When a fold is applied **perpendicular** to the grain of the paper, the resulting fold may have a ragged appearance. A ragged fold is produced

**creep** – thickness of paper causes the inner pages to stick out further than the outer pages
**allowance** – the amount of something that is permitted
**parallel** – a line that is always the same distance from another line
**perpendicular** – at right angles to another line

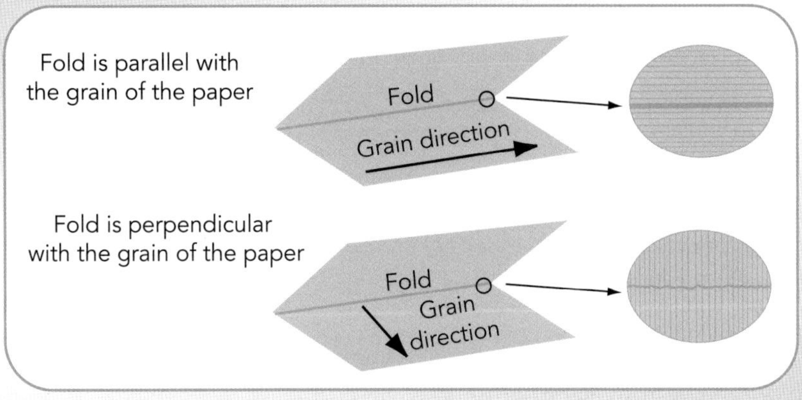

FIG 3.9: Effect of folding parallel or perpendicular to the paper grain

because all of the fibres are folded at one time, creating resistance to the folding action. The ragged appearance of the fold is especially noticeable when folding heavy paper.

# Types of binding

Binding is the process of fastening loose sheets of paper together.

## Saddle stitching

A simple way of assembling a small booklet or magazine with a wire stitch (staple) through the fold. Printers call it saddle stitching, not stapling.

## Thread section sewing

This is usually found in books as it is one of the most durable binding processes. As the name suggests, it is a technique where the pages are held in place by literally sewing them together.

## Wire binding

This is a process of securing the pages with a coil of wire clamped through pre-punched holes. Sometimes the wire is replaced with plastic, and the name given to this technique is comb binding. Calendars often use this type of binding.

## Perfect binding

A binding technique whereby single sheets are stacked together, the binding edge is **ground** to create a rough surface, and glue is applied. A cover is then wrapped around the pages. Many magazines and most paperback books are perfect bound. A wide range of glues is used, including cold and hot glue.

# Range of paper sizes

For many printed products the overall size of the finished product will be influenced by the number of pages that can be **economically** printed together on one sheet.

In the UK the common system of paper sizes is the ISO standard. This includes the A series that you are familiar with – ranging from A0 at 841 x 1189 mm to A8 at 52 x 74 mm. There are also a B and C series. Two other series, RA and SRA, are used by printers. They are slightly larger than the A series to provide for grip, trim and bleed.

It is not only the size of the paper that is considered: paper type and quality (or grade) are also essential elements. The choice of paper will be dependent on the final product.

The presses at Westway use silk, gloss and matt **stocks** of paper with a weight range from 80–90 g/m$^2$ up to 350 g/m$^2$.

**ground** – roughened by grinding

**economically** – in a way that uses no more of something than is necessary

**stock** – general term for any paper or board which is used as a printed surface

You may also see the units g/m$^2$ as gsm (grams per square metre). It is simply an indication of the weight of the paper (or other stock). For example, a typical photocopier paper would be 80 g/m$^2$, a good letterhead paper might be 100 g/m$^2$ and a postcard would be about 250 g/m$^2$.

### How is the paper fed to the press?

A **web** printing machine is one that accepts the paper on a large roll (web). These are very fast presses and are only economic for long-run and high-volume work. Most people have seen film of newspapers being printed – this is a web process. The majority of magazines you find in the newsagents have been printed by web.

The sheet-fed press prints on individual sheets of paper.

## Modern technology

### Digital printing

Digital printing is printing technology that can produce printed sheets directly from a computer file, without going through some intermediate **medium** such as a film negative or an intermediate machine such as a plate-making machine. Benefits of this process include faster **turnaround** times, lowered production costs, and the ability to personalise documents. It is frequently used for on-demand or short-run colour printing.

**medium** – a method or way of expressing something

**turnaround** – the time needed to complete a task

There are two main kinds of equipment used for digital printing:

- direct imaging press – these are based on an offset, or conventional, printing method
- digital colour printer – laser printer, inkjet printer.

This print technology is used for:

- on-demand printing
- variable data printing
- web-to-print.

**On-demand**, or print-on-demand, allows for small quantities of printing to be done as needed, with turnarounds of a few hours or less. This printing is perfect for companies, organisations and individuals who need to constantly update their printed material. A digital colour printer is usually used for on-demand printing.

**Variable data printing**, or variable image printing, is a digital print run where each printed page is somewhat different, with the variations usually determined by relating the page content to customer information in a database.

For example, a set of personalised letters, each with the same basic layout, can be printed with a different name and address on each letter. Variable data printing is mainly used for direct marketing, customer relationship management and advertising.

The technique is a direct outgrowth of digital printing, which utilises computer databases and digital presses to create high-quality, full-colour documents with a look and feel comparable to conventional offset printing.

Variable data printing enables the mass customisation of documents via digital print technology, as opposed to the mass production of a single document using offset lithography. Instead of producing 10,000 copies of a single document, delivering a single message to 10,000 customers, variable data printing can print 10,000 unique documents with customised messages for each customer.

**Web-to-print** digital printing, or web-enabled printing, allows for direct mail pieces to be customised and personalised online. Clients and customers can choose images, such as photographs, to include in materials such as brochures or greetings cards. A proof is shown online, and when the piece is ready, one click sends it to the print supplier for rapid delivery.

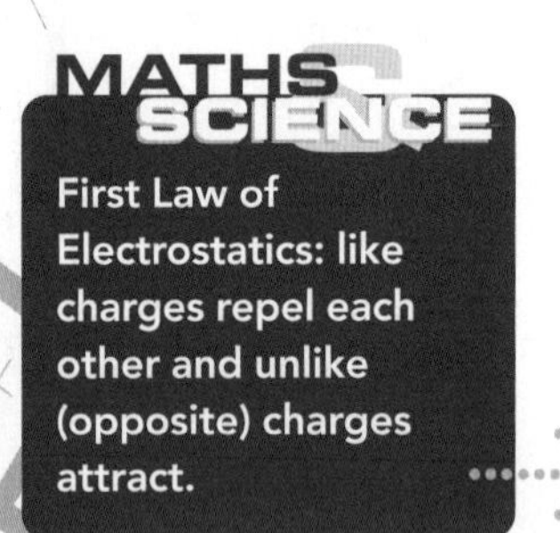


First Law of Electrostatics: like charges repel each other and unlike (opposite) charges attract.

## Digital printing devices

**Electrostatic printing**: Photocopiers and laser printers use the concept of electrostatic printing. Basic to the electrostatic printing technique is the

**FIG 3.10:** Photocopiers use the concept of electrostatic printing

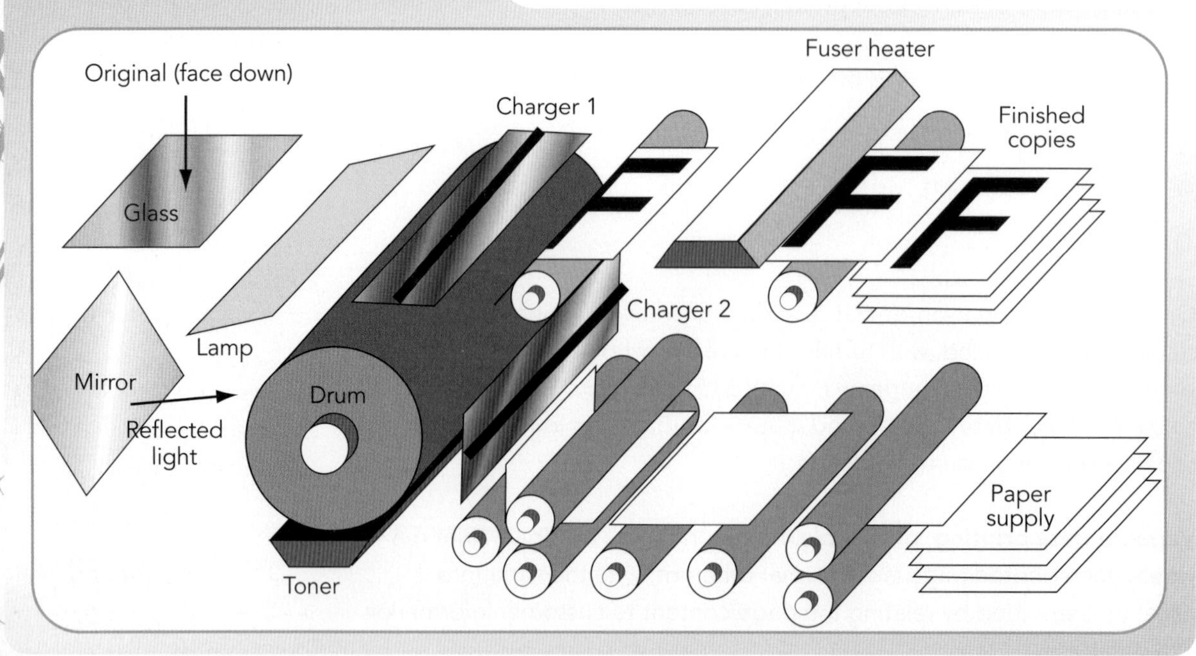

fact that particles having opposite electrical charges will attract each other.

In electrostatic printing, the surface of a selenium-coated drum is given a positive electrical charge.

The image is scanned and projected onto the drum. The image areas on the original do not reflect light, and that produces a positive electrical charge. Since positive charges attract, they pull the **toner**, which carries a negative charge, from the toner cartridge, onto the areas that carry a positive charge – the image areas. Heat is then applied, which **fuses** the particles of toner to the blank sheets of paper, thereby reproducing the image on the paper.

**MATHS & SCIENCE**
Some materials become better conductors when exposed to light – this is true of selenium.

**Inkjet printers**: One of the most popular and affordable printers available today is the inkjet printer.

An inkjet printer is any printer that fires extremely small droplets of ink onto paper to create an image. The dots of ink are:

- extremely small (usually between 50 and 60 **microns** in diameter)
- the dots are positioned very precisely, with resolutions of up to 1440 x 720 dots per inch (dpi).

**MATHS & SCIENCE**
A micron is one millionth of a metre or one thousandth of a millimetre.

There are two main technologies currently used to fire the ink: the **thermal bubble** and the **piezoelectric**.

With the thermal bubble or bubblejet printer, **resistors** are used to create heat, which then creates a bubble in the ink. The bubble expands and forces ink out from the nozzle onto the paper. When the bubble 'pops' (collapses), more ink is drawn into the print head from the cartridge. On average, a bubblejet printer will have a range of three hundred to six hundred nozzles.

The piezoelectric printer uses small crystals in the nozzles that vibrate under the influence of an electric current. This vibration pushes ink out of the nozzle and subsequently draws more ink in to replace the ink sprayed out. The drops of ink that come from the piezoelectric type printer are significantly smaller than those of the bubblejet printer, allowing for greater control over the image quality.

**toner** – dry ink powder used in printers, fax machines and photocopiers
**fuse** – melt a material with intense heat so as to join it with something else
**resistor** – a device that has resistance to the passage of an electric current

# The process

## Introduction

One of the most valued customers at Wyndeham Westway is the British Heart Foundation (BHF), who supply approximately 30 jobs per month. These range from brochures, leaflets and direct mail to educational materials, medical fact sheets and magazines.

One magazine, as illustrated in the film, is entitled *Saving Lives. Saving Lives* is an A4, 36-page magazine with a print run of 10,000 copies per issue.

**FIG 3.11:** Flow chart of the process

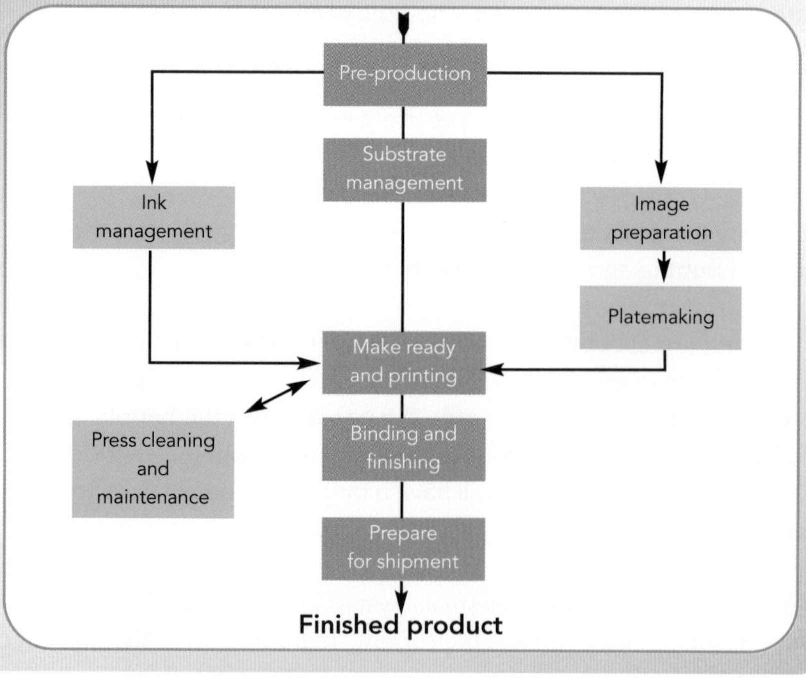

## Planning

### The brief

The initial brief, or specification, is received from the client via email or by telephone. The specification will include details on the:

- number of copies required
- paper weight
- finished size
- number of colours required

- finishing, e.g. saddle stitched, perfect bound, laminated, etc.
- required delivery date.

Once the specification has been received, it is necessary to decide on which factory, within the Wyndeham Group, is best suited to produce that job. Factors to consider will include the size of the job and the length of turnaround required.

The initial brief is then forwarded to the estimating department, who may need to clarify specific details, including finishing requirements and elements that may be specific to that job. They put this information into their **Management Information System** (MIS), which generates a price for the job – a quote.

The costs are dependent on the price of paper and inks at the time, and also the time of year (prices are generally higher at busier times within the print facility).

Print is a business with a high level of **capital expenditure.** Therefore, the price at which a job is sold to customers also needs to cover all of the factory **overheads,** e.g. the machinery (upgrade and maintenance costs), labour costs, heat and light, etc.

Due to the fluctuating nature of the prices of raw materials and the overheads of the print facility itself, it is imperative that all quotes are prepared from scratch every time they come through.

Once the client accepts the job, an **order confirmation** is generated. A **job bag** is then created, which details every piece of information that will be used to create and complete the job.

The job will then be scheduled in for production using a computerised system.

**capital expenditure** – amount used during a particular period to acquire or improve long-term assets

**overheads** – the cost or expense incurred in the upkeep of running a business

**job bag** – set of instructions to departments which enables them to complete the job to the customer's requirements

**entities** – a thing with distinct and independent existence

# Production and testing

## Introduction

The production and testing processes involved in print production are not separate **entities**. It would be foolish to undertake a whole print run and only then check the quality and accuracy of the finished product. Testing, or in this case the checking of the product, is performed at an early stage, and it is only when this check has ensured that the product is the one requested by the client that the full print run is authorised.

## The process

Once the schedule has been agreed with production it is imperative that the artwork is received from the client.

Today artwork exists almost wholly in electronic form. Photographs and illustrations are input to the computer using a scanner. All the elements are assembled using page layout software.

The client can supply the artwork files to Wyndeham Westway in a number of ways:

- CD – through the post
- via the internet – using the Wyndeham site they can securely upload the files, which are then downloaded in the Westway artwork studio.

**FTP** – file-transfer protocol
**ISP** – internet service provider

These days, customers regularly upload their files via the internet. Files can be submitted by **FTP**, email and the internet. FTP is probably the most appropriate method for handling large files. Email is difficult to use, as print files are usually large in size, many **ISPs** restrict the size of the file attachments, and email delivery speeds are often unreliable.

NUTS & BOLTS

**To find out more about page layout software go to http://desktoppub.about.com/cs/software/p/sw_pagelayout.htm**

The artwork files are received by Westway in the pre-press studio. **Pre-press** is basically the process of converting artwork files into a format suitable for print.

The pre-press department will **pre-flight** the artwork, which means they will check the files to ensure:

- usability
- they are print ready
- the images are high resolution
- the document supplied is in the same format as the specification raised for the job bag.

**Imposition** will also be undertaken at the pre-press stage. This is the process of arranging the pages so that when the sheets are printed and folded for binding, the pages will be in the correct order.

Once they are satisfied with the files they will produce a press proof. A **proof** is simply a test print to show what the finished product will look like, and provides a method of checking for errors prior to printing an order.

Proofs can be made in a variety of different ways. The simplest form is a colour laser or inkjet print, which can create a rough impression. Increasingly popular are digital proofing systems, which are essentially colour printers with a very high resolution which make use of colour management techniques for their accuracy.

Once the proofs have been approved by the customer and the customer service executive, they are returned to the pre-press department for plate making. After the plates have been made they are taken to the press for the print run.

A **print minder** – the person who actually runs the press – puts the plates in the machine and undertakes a process known as **making ready**. This entails getting a press ready for a print run. The minder will check the **registration** of the print item, i.e. he or she will check the colours to see how they sit on top of each other on the paper.

HEALTH & SAFETY

Up to 40% of all machinery-related accidents in the print industry occur on printing presses.

Checking the registration of the printed item

When printing with two or more colours it is necessary to align the different plates. This is known as **register**. On the edges of an untrimmed sheet there are small target shapes called register marks which are used for accurate positioning. If one of the plates is slightly out of register, the image will appear blurred.

After completing all the checks the client will be called in to complete a final check – **press pass** – before signing off the job, enabling the rest of the run to proceed.

HEALTH & SAFETY

Transportation and loading of reels and sheets can cause back injuries. Lifting and handling aids must be used.

The specific paper chosen for the end product will be loaded onto the press. The presses at Westway are sheet-fed, which means the printing press uses large pre-cut individual sheets of paper, as opposed to continuous rolls of paper.

Once printed, the material will be taken to the folding and binding line for finishing. At Wyndeham Westway the stitching and binding lines are extremely fast and will allow in the region of 8–12,000 copies to be collated and stitched in an hour.

Binding line

In the film you hear that for the *Saving Lives* magazine there are four elements: two 16-page sections and two 4-page sections. A **section**, in the printing industry, is a folded sheet which is assembled with others to make up a book. For example, an A2 sheet will provide a section of eight A4 pages when folded twice. A 20-page booklet would therefore require two 8-page sections and one 4-page section.

The final stage of the process is the dispatch of the product.

Binding line

# The career path

## PERSONAL PROFILE: Nichola King

**Qualifications**
3 A levels
11 GCSEs

**Training**
Leadership and personal development training.

**Professional experience**

| | | |
|---|---|---|
| *New Business Development Manager* | *2004 – Present* | *Wyndeham Press Group Plc* |

- Responsible for new business sales.
- Sold £1.8 million in year one of new business.
- Development role in strategy and implementation of new business sales across all group businesses.

| | | |
|---|---|---|
| *Sales Manager and Business Development* | *2000–2004* | *Miter Press Ltd* |

- Directly responsible for managing the sales team.
- Involvement in sales training and team development.
- Marketing to FTSE 250 companies, agencies and print management companies.

| | | |
|---|---|---|
| *Director* | *1998–2000* | *CMYK Ltd* |

- Legal director with joint ownership of the company.
- Overall commercial responsibility for print management services: sales, design, typesetting, reprographics and print.
- Company was profitable after one year.

| | | |
|---|---|---|
| *Sales Manager* | *1998–1999* | *Computer Publishing Ltd* |

- Marketing to IT specialists, recruitment agencies and marketing agencies.
- Responsible for expanding existing business.
- Increased magazine pagination from 8pp to 32pp fortnightly.

| | | |
|---|---|---|
| *Field Sales Representative* | *1995–1998* | *Home Counties Newspapers Ltd* |

- Managed initial sponsorship of a business awards ceremony – now an annual event.
- Proofing advertising sections of newspaper.

**To find out more about careers in the print industry go to www.britishprint.com and www.prospects.ac.uk**

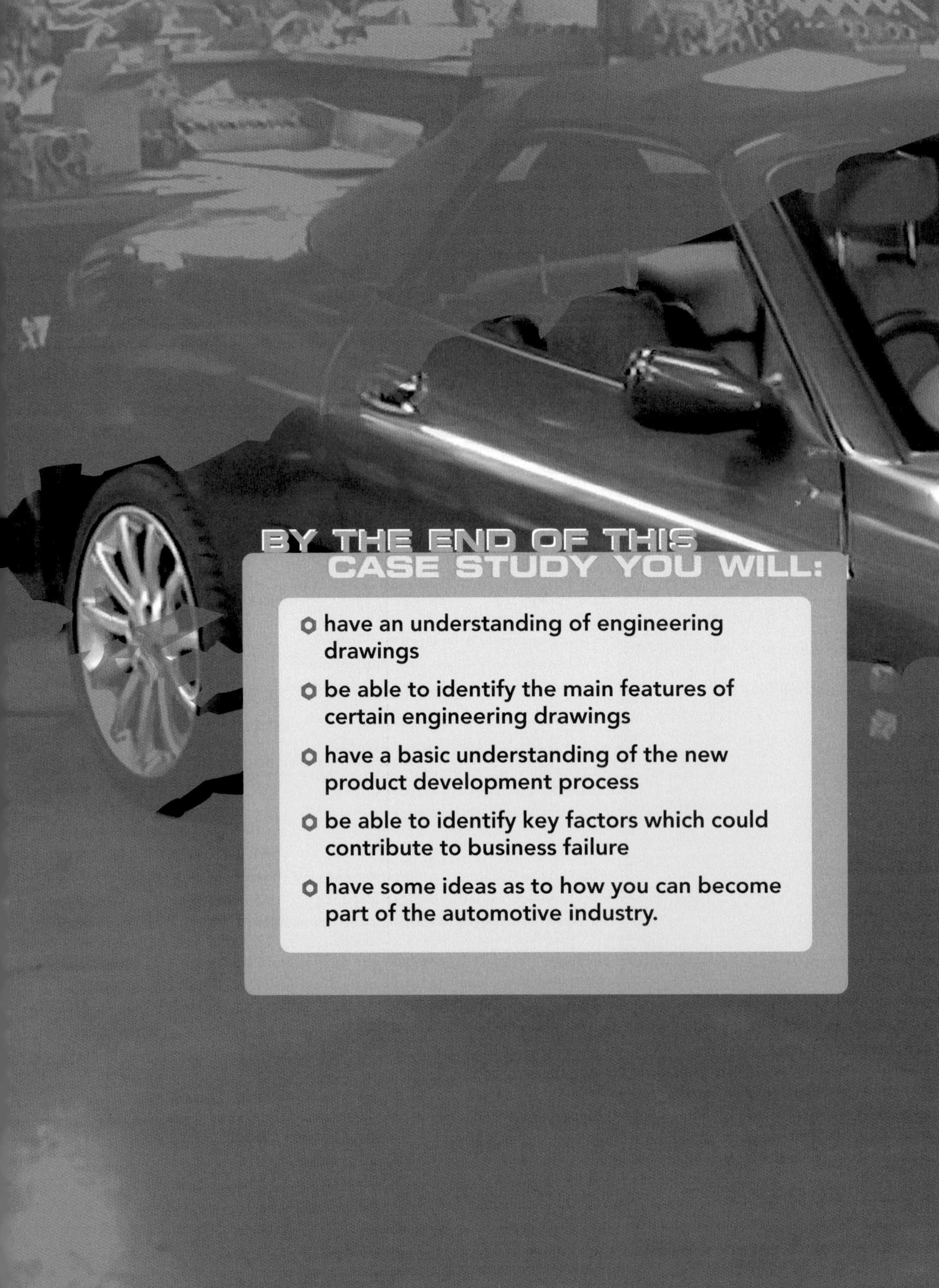

## BY THE END OF THIS CASE STUDY YOU WILL:

- have an understanding of engineering drawings
- be able to identify the main features of certain engineering drawings
- have a basic understanding of the new product development process
- be able to identify key factors which could contribute to business failure
- have some ideas as to how you can become part of the automotive industry.

# SMALL ENGINEERING COMPANY: MIKE SATUR

## Case study overview

This case study introduces you to a small business enterprise operating within the automotive industry.

Over the following pages you will become familiar with a real company that designs and develops products for the MG car.

After looking at a range of engineering drawings, you will consider the company's processes from initial idea to final product. This will help you to understand and appreciate that whatever the size of a company, new product development is a vital part of business growth.

Finally, you will be introduced to the owner of the company who will describe the route he took to his current position.

# The sector

## The engineering industry

The engineering sector is made up of many industries – energy, transport, electronics and telecommunications, defence, the **built environment**, metals and medical equipment.

In fact, the engineering sector is huge. There are an estimated 800,000 engineers in a sector employing roughly 1.7 million people in total. At the end of 2005, the total number of Engineering Council UK registered engineers stood at 243,077. Of these, 188,367 were **chartered engineers**, 41,603 **incorporated engineers** and 13,107 engineering technicians.

**built environment** – refers to man-made surroundings, e.g. buildings, roads
**chartered engineer** – highest level achievable within the engineering profession
**incorporated engineer** – performs the technical duties on a project
**value added** – increase in worth in the economy as a result of a particular activity
**GDP** – gross domestic product: the total value of goods produced and services provided in a country during one year

## The automotive industry

The automotive manufacturing sector contributes around £10 billion **value added** to the economy, and accounts for 1.1 per cent of **GDP**. Some 210,000 people are employed in the design and manufacture of vehicles and components, and a further 570,000 in the motor trades which supply, service and repair vehicles in the UK.

## Small business enterprise

The most recent government figures show that there were around 4.3 million business enterprises in the UK at the beginning of 2005, an increase of 59,000 on the year before. Of all the firms, around 99 per cent were small businesses employing less than 50 people.

These small and medium-sized enterprises (SMEs) employ around 12 million people and are vital to the UK economy.

**NUTS & BOLTS**

**To find out more about the automotive industry go to www.autoindustry.co.uk**

# The company

## Background

Mike Satur is the company director of a specialist small company, based in the north of England, which designs and develops particular products for the MG car. These products are aimed at improving the **roadholding**, braking, performance and reliability of the MG, as well as enhancing **aesthetic** appeal.

With over a quarter of a century in MG ownership and involvement, Mike Satur offers a range of original quality accessories specifically tailored for the MG owner. The business has remained competitive, as it has responded to the many changes in the market since the 1960s.

This small enterprise is a success through the dedication and thoroughness of the employees and the team focus on getting things right first time. Mike Satur is not a '**run-of-the-mill**' garage; this innovative motor engineering company provides a unique, specialist service to its customers.

**roadholding** – the ability of a vehicle to remain stable when moving
**aesthetic** – concerned with beauty
**run-of-the-mill** – ordinary or not special in any way
**stock exchange** – place where stocks, bonds or other securities are bought and sold

## Modern-day concerns

40,000 new businesses start up each year in the UK. However, as many as one-fifth of these businesses cease trading within their first 12 months. Furthermore, a significant proportion will not exist beyond the five-year mark.

There are many reasons that a business can fail, but one of the main factors is that of money, or, more precisely, a lack of it when it's most needed. Manufacturing, retail, and transport and communication businesses can also be affected by natural disasters in countries where materials and products are sourced from, or a weak pound on the **stock exchange**.

The challenge for small businesses is ensuring that they are able to cater for ever-changing market conditions.

**To find out more about Mike Satur go to www.mikesatur.co.uk**

# The product

## Introduction

The products designed and developed by Mike Satur are many and varied, as dictated by individual customer demands and **fluctuating** market conditions. This is the inevitable result when a business caters for such a **niche** market, providing a **one-off** service. As such, for this case study it is more appropriate to consider the challenges faced by small business enterprises. Firstly, however, we shall consider the starting point for Mike Satur's product design, that of the engineering drawing.

**fluctuate** – to change or vary
**niche** – a specialised but profitable segment of the market
**one-off** – done only once
**nomenclature** – a system for naming things
**typeface** – style and size of printed letters
**geometric** – describes a pattern or arrangement that is made up of shapes
**template** – something that serves as a model for others to copy

## Engineering drawings

One of the best ways of communicating ideas is through some form of picture or drawing. This is especially true for the engineer.

An engineering drawing is a type of drawing that is technical in nature, and is used to fully and clearly define requirements for engineered items. These drawings are usually created in accordance with standardised conventions for layout, **nomenclature**, appearance (such as **typeface** and line style), size, etc.

The purpose of the drawing is to accurately capture all the **geometric** features of a product or a component, conveying all the required information that will allow a manufacturer to produce that component or product.

## Drawing layouts

Engineering drawings are usually drawn on pages with a set **template**. This is to enable the drawings to be referred to, stored and, consequently, retrieved easily.

You may wonder why drawings are not all stored on a computer. If you store drawings only on a computer you would also need a computer to read them. This is impractical in, for example, a garage or workshop where you are following a drawing whilst working on a car. It is also impracticable in many other manufacturing situations and, as such, paper drawings will always have a place in engineering.

While drawing templates vary from company to company, certain key details are vital and are always included, usually within an information box.

Where an engineering component is complicated there will be a number of different drawings, one showing each part of the component. The drawing numbers will follow a sequence and will be included in a list known as a **parts list**. The parts list can either be a separate document or be included on the master drawing sheet.

You will notice that the layout drawing template shown also has reference numbers and letters, like on a map. This makes it easier to reference a section of the drawing when discussing with a fitter, client or manufacturer a part or feature of the design over the phone.

**MATHS & SCIENCE**
Metric units of length are millimetres.
10 mm = 1 cm.
100 cm = 1 m.

**FIG 4.1:** Example of a drawing template

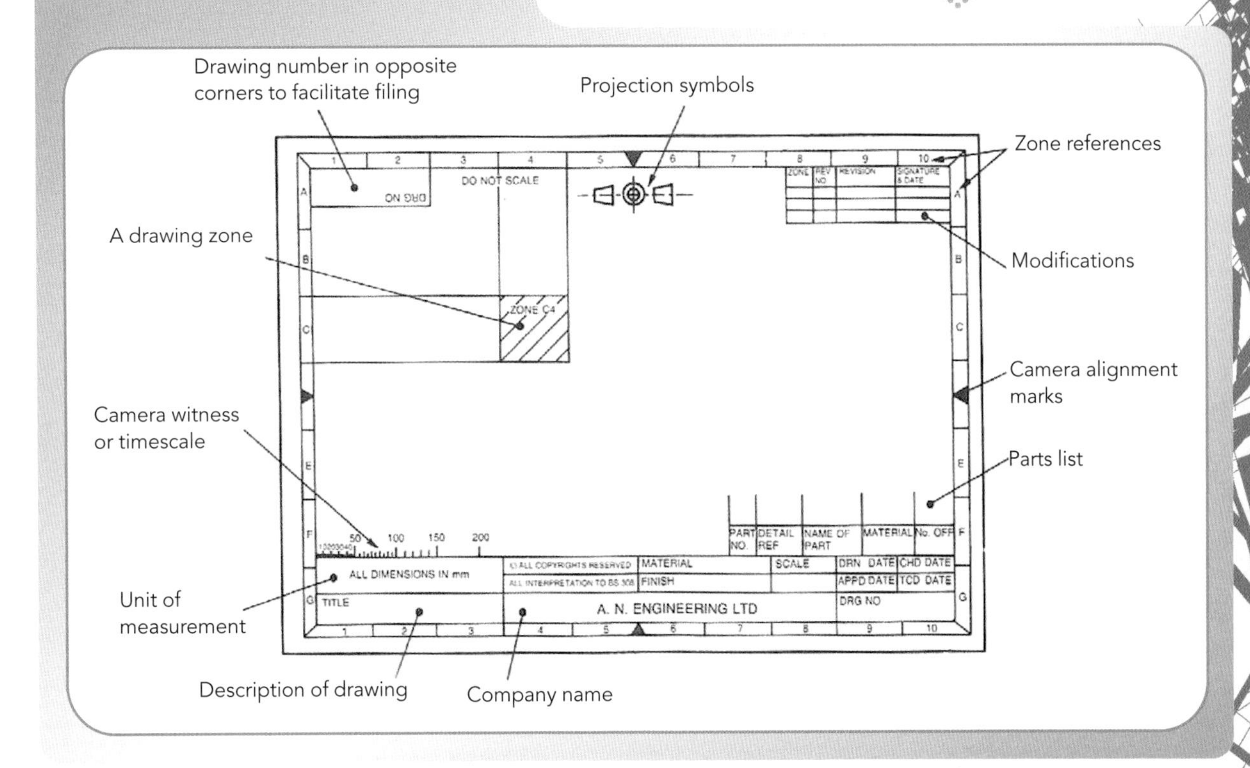


# Sketching

Designs are usually first produced as a freehand sketch. Whilst computer simulation enables a designer to get a feel for the three dimensions of a product, freehand sketching is normally the way a product is first realised outside the engineer's head.

Representing three dimensions on a flat piece of paper is a very important skill for engineers, enabling them to communicate their ideas clearly to other people. Freehand sketches are usually drawn using one of three different techniques. These are perspective, isometric or oblique.

**MATHS & SCIENCE**
Scale. 1:1 means the drawing is the same size as the object. 1:10 means 1 cm on the drawing represents 10 cm on the object.

**MATHS & SCIENCE**
Three-dimensional objects have three dimensions – length, breadth and depth.

## Perspective

All objects we look at have **perspective**. Objects closer to us are bigger

**perspective** – appearance of viewed objects with regard to their relative position, distance from the viewer, etc.
**vanishing point** – point at which receding parallel lines viewed in perspective appear to come together
**horizon** – the line at the most distant place which you can see

than objects further away. Or, as objects get further away they seem to 'vanish' into the distance.

In perspective drawings, lines slope to one or more **vanishing points** (VP). Perspective drawing can use one, two or three vanishing points.

One-point perspective is normally used for room interiors. Two-point perspective is used for developing ideas in three dimensions, and three-point perspective is often used for drawings of tall buildings.

FIG 4.2: Example of a two-point perspective drawing

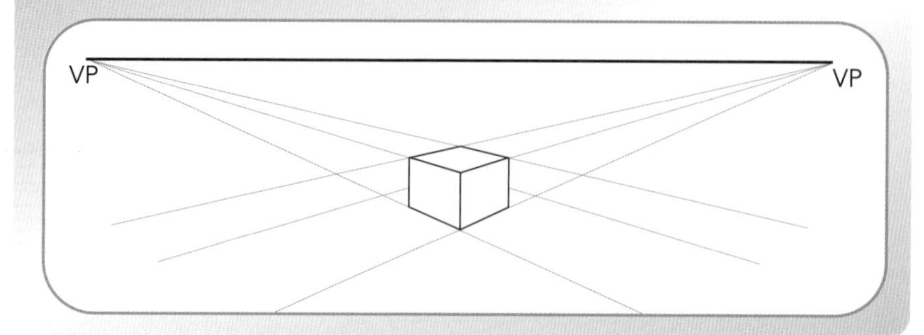


The diagram above shows a two-point perspective. This is a more useful drawing system than the one-point perspective as the object has a more natural look.

In two-point perspective the sides of the object vanish to one of two vanishing points on the **horizon**. Vertical lines in the object have no perspective applied to them.

## Isometric

Isometric is a mathematical method of constructing a three-dimensional object without using perspective.

FIG 4.3: Isometric drawing

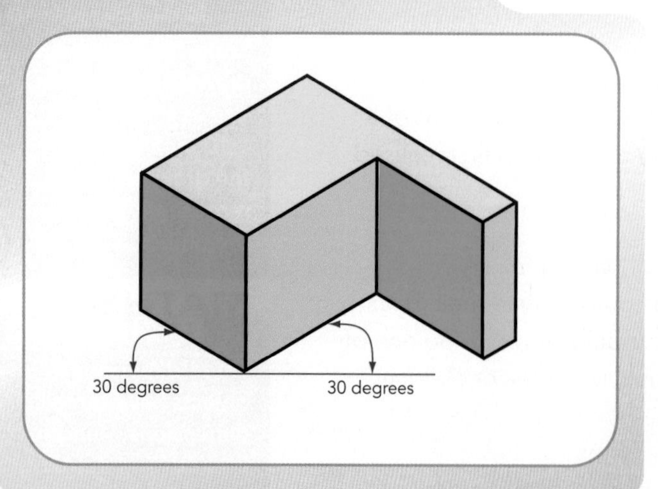


FIG 4.4: Oblique drawing

An isometric drawing shows two sides of the object and the top or bottom of the object. All vertical lines are drawn vertically, but all horizontal lines are drawn at 30 degrees to the horizontal.

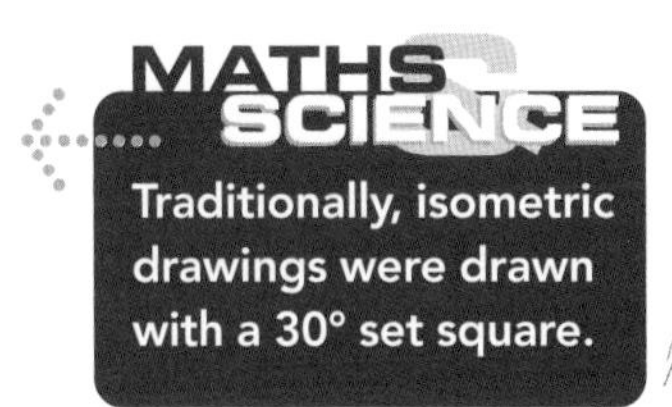


Isometric drawings are normally produced with drawing equipment to ensure accuracy.

Isometric drawings can be drawn to scale and are an easy method of constructing a reasonable three-dimensional image.

### Oblique

Oblique drawing is quite a simple technique compared to isometric or even perspective drawing, and, indeed, it is the easiest to master.

When using oblique, the side of the object you are looking at is drawn in two dimensions, i.e. flat. The other sides are drawn in at 45 degrees but, instead of drawing the sides full-size, they are only drawn with half the depth, creating 'forced depth' which adds an element of realism to the object.

**MATHS & SCIENCE**

**To produce an accurate 45° angle, you can bisect a right angle (90°).**

## Working drawings

### Orthographic

Any engineering drawing should show everything. A complete understanding of the object should be possible from the drawing. In order to get a more complete view of an object, an orthographic projection may be used.

Orthographic projection shows complex objects by giving a two-dimensional drawing of each side, to show the main features.

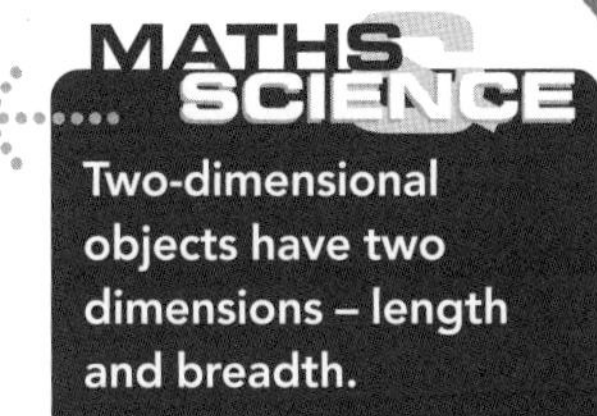


An orthographic projection shows the object as it looks from the front, right, left, top, bottom or back. These drawings are typically positioned relative to each other according to the rules of either **first angle** or **third angle projection**.

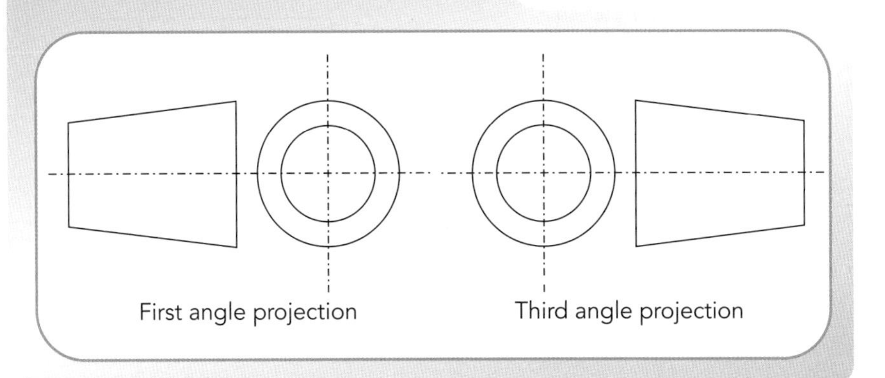


FIG 4.5: Angle of projection symbols

Third angle projection is as if the object were a box to be unfolded. If we unfold the box so that the front view is in the centre of the two arms, then the top view is above it, the bottom view is below it, the left view is to the left, and the right view is to the right.

Both first angle and third angle projection result in the same views; the difference between them is the arrangement of these views around the object.

When drawing in first or third angle projection, a symbol is used to show which angle of projection has been used. See Figure 4.5.

**FIG 4.6:** Orthographic projection

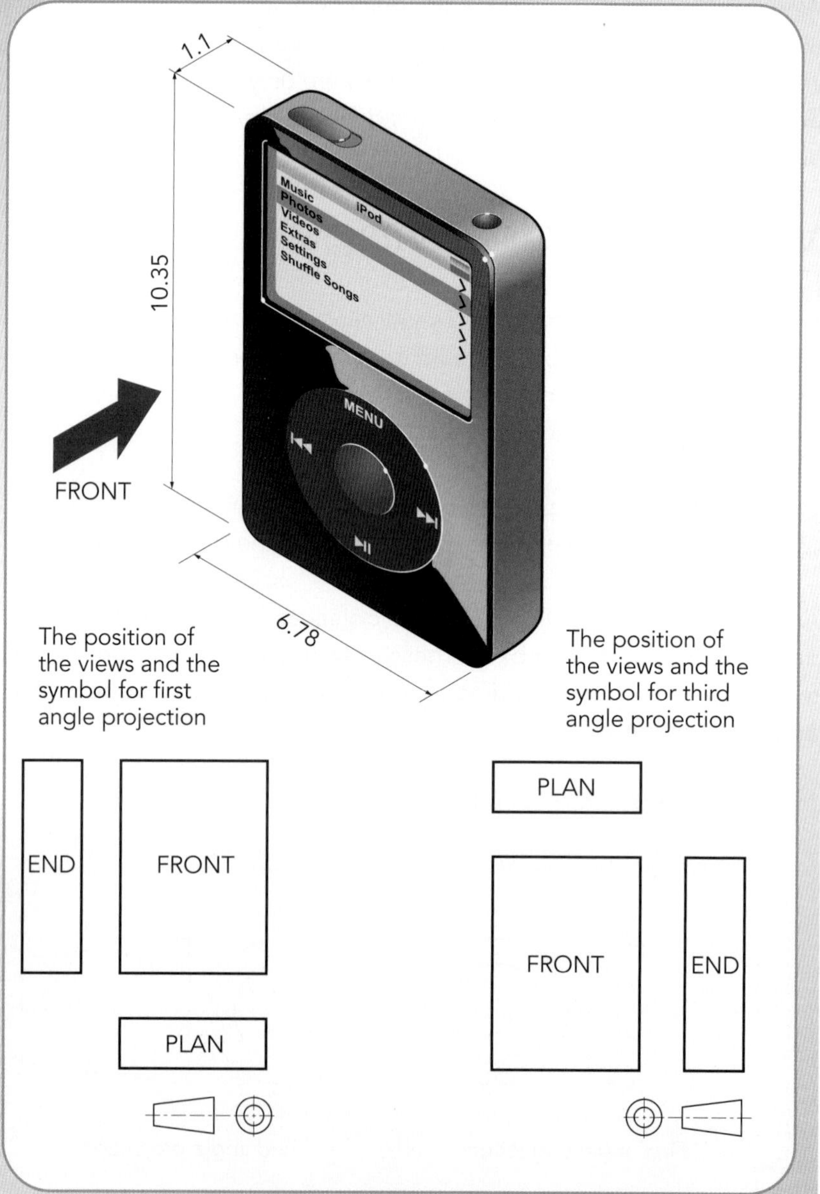

1.1
10.35
6.78
FRONT

The position of the views and the symbol for first angle projection

END
FRONT
PLAN

The position of the views and the symbol for third angle projection

PLAN
FRONT
END

In Figure 4.7, shading has been added to help identify the different views.

**FIG 4.7:** Shading is used to identify the different views

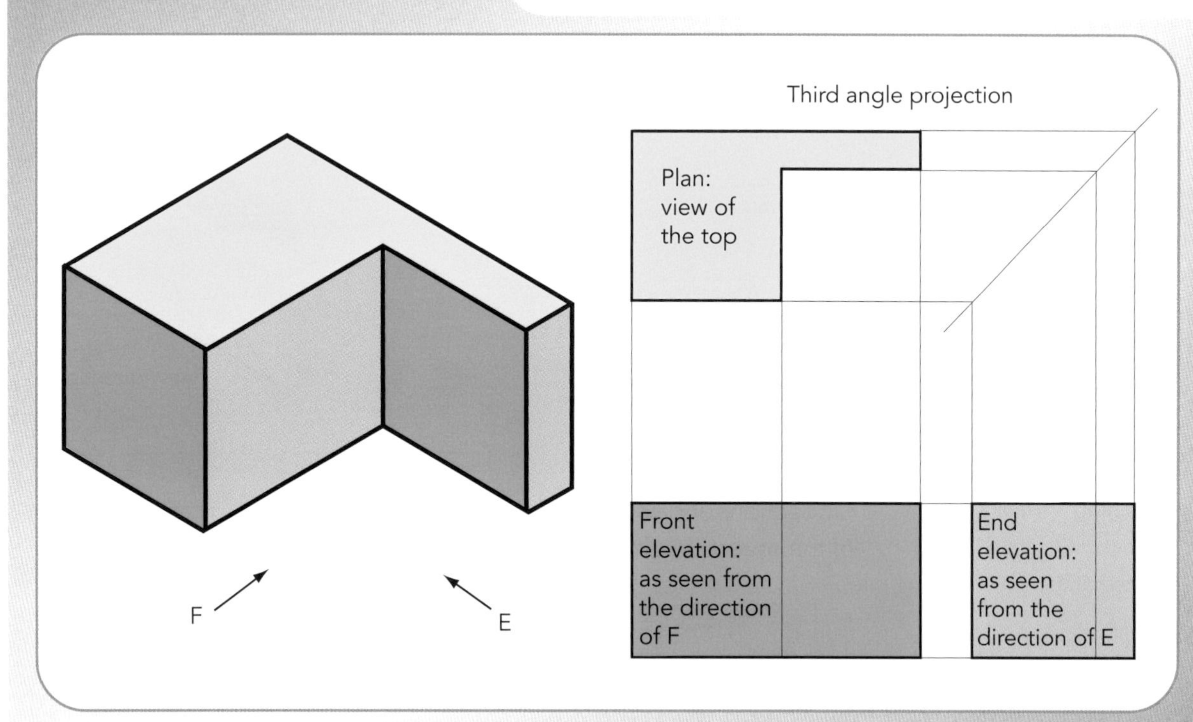


A **plan drawing** of an object is the two-dimensional horizontal view that is seen when the object is looked at from a position above the object and looking straight down.

An **elevation drawing** of an object is the two-dimensional vertical view seen when the object is looked at from a position to one side of the object and looking straight at it.

# Sections

Where an object has complex internal shapes it is common to show a section. Sections and sectional views are used to show hidden detail more clearly and to give extra structural information.

A section is simply the object shown in one view as if it had been cut using a saw. A section is a view of no thickness, which shows the outline of the object at the cutting **plane** – the position of the imaginary cut.

**plane** – imaginary flat surface through objects

The cutting plane, or section plane, is sometimes represented by a line consisting of long and short dashes.

Figure 4.8 shows a sectional view, and how a cutting plane works.

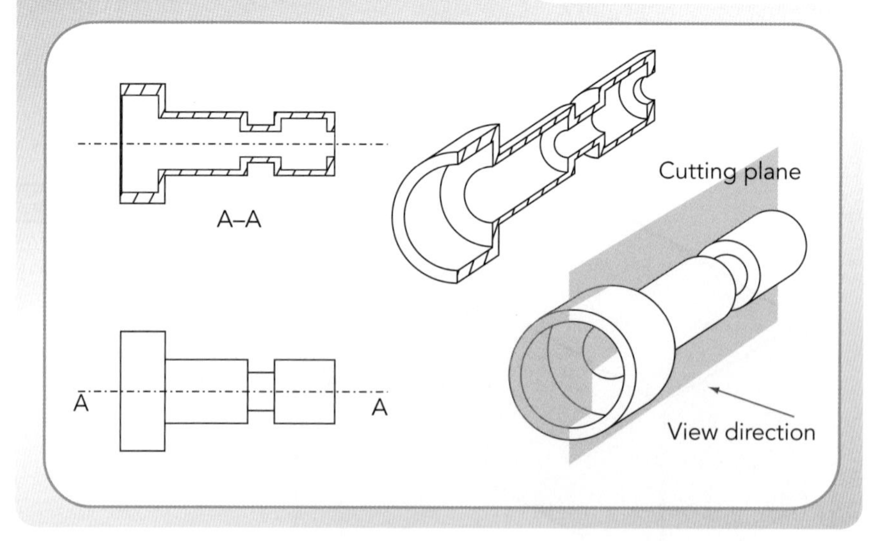


FIG 4.8: Sectional view and how a cutting plane works

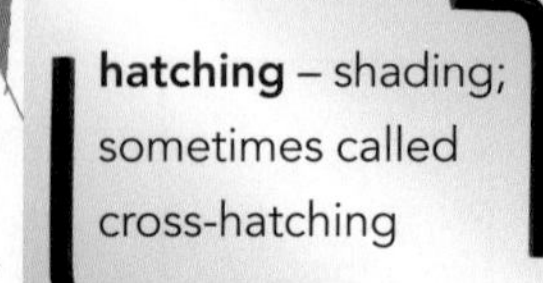
**hatching** – shading; sometimes called cross-hatching

On sections and sectional views, **hatching** is used to indicate those parts of the object that have been cut through.

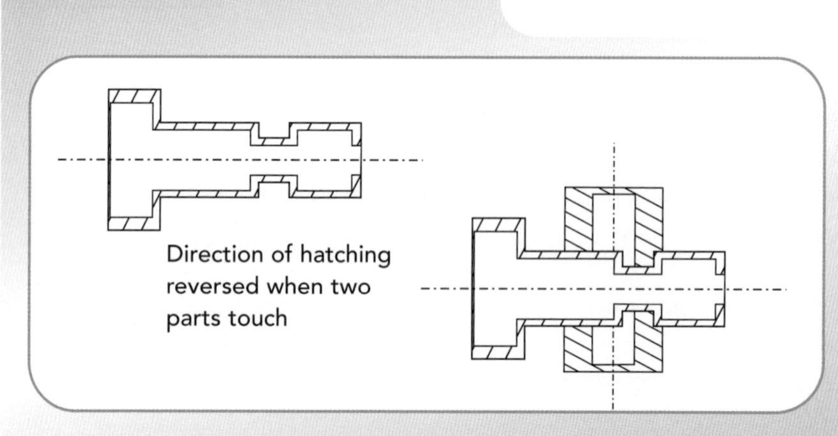


FIG 4.9: Hatching

The hatching is drawn with thin lines at 45 degrees and spaced 4 mm apart.

If two parts of an object are touching, the direction of the hatching is reversed.

## Auxiliary views

Another way of showing detail is by drawing what are known as auxiliary views. Auxiliary views are additional views that are used in orthographic projection.

An auxiliary view is a projection of the object from a different angle. Auxiliary views are important when, like in the seat belt holder shown in the film, parts of the component are at an angle, or the designer wishes to show the object from a different angle.

The auxiliary view gives a deeper insight into the actual shape of the object.

### Exploded views

Often in instructional manuals, particularly in the motor industry, another type of view is used, which is called an exploded view.

An exploded view is a representative picture or diagram that shows the components of an object slightly separated by distance, or suspended in surrounding space in the case of a three-dimensional exploded diagram.

Exploded diagrams are common in descriptive manuals showing part placement, or parts contained in an assembly or sub-assembly. Usually such diagrams have the part identification number and a label indicating which part fills the particular position in the diagram.

**NUTS & BOLTS**

**To see examples of exploded view diagrams go to www.mossmotors.com/SiteGraphics/CataloguePDFs/MGTC-0702.pdf**

## Assembly drawings

An assembly drawing, sometimes known as a general arrangement drawing, shows how the various parts of a product are to be assembled, i.e. how they fit together.

Each part is given a unique number, and a table of parts is added to the drawing to identify each part. For example, a parts table could be constructed as follows:

| Item No. | Description | Qty | Material | Remarks |
|---|---|---|---|---|
| | | | | |
| | | | | |
| | | | | |
| | | | | |
| | | | | |
| | | | | |

**NUTS & BOLTS**

**To find out more about engineering drawings go to http://ider.herts.ac.uk/school/courseware/graphics/engineering_drawing**

# The process

## Planning

In business and engineering, **new product development** is the term used to describe the complete process of bringing a new product or service to market. Product development is important as it controls the costs and time taken to develop a new product.

Within the new product development process, there are two parallel paths of consideration:

- idea generation → product design → detail engineering
- market research → marketing analysis.

Some 'new' products, as in the case of Mike Satur engineering, are simply a result of modifying existing products.

Initial research into the potential market needs to be carried out, which may include producing reports on the context for the product's use and consumer trends. Future developments for the product may also be included at this stage.

For small or large companies to succeed in product development, the development is divided into a number of stages. However, this is not a fixed process and may vary depending on the different types of companies.

Where more than one person is involved, the process usually starts with a briefing; this is a meeting to produce a common understanding of what is required. The brief can be complex and is often backed up by research.

Briefing usually covers three main areas:

- marketing – describes the product, what it does and what other similar products already exist
- product specification – covers performance, cost, intended manufacture, and standards (e.g. health and safety) that need to be considered
- commercial – relates to sales and distribution, including sales targets and **forecasts**.

For Mike Satur engineering, understanding what its customers want is an essential part of the initial product development process.

**forecast** – prediction or estimate of a future financial trend

In the film, Mike Satur is clear about the market for his products – he is meeting the needs of a specialised market, that of MG owners. The market is **buoyant** because people who own MGs are enthusiasts. He stresses the importance of understanding the customer's needs and catering for them directly.

**buoyant** – involving or engaged in successful trade or activity

# What can go wrong?

## A bad idea to begin with

It is sometimes the case that the 'unique' business idea is not so unique after all and that the market is already well served, or that there is less demand than originally thought.

## Not knowing the market

Knowing who your customers are and what they actually want, rather than what you believe they want, is one of the keys to business survival. Failure to keep in touch with customers through implementation of a practical marketing plan and failing to keep up with their changing wants and needs is a recipe for disaster.

## Failing to change

No business exists in a bubble. Failing to take note of changes in the business environment can lead to problems. What the competitors are doing, changes in technology and **best practice**, as well as changing patterns in the customer's buying patterns and tastes, need constant monitoring.

## Design

Coming up with and developing new ideas to keep the company competitive is one of the problems outlined by Mike Satur. For small businesses, particularly those providing specialist services, this is a critical factor in determining future success. Mike Satur's success was partly due to his ability to find new ideas and develop new products because of his own enthusiasm and love for the MG.

The process of the idea **genesis** can be made internally or come from outside inputs, e.g. from a supplier offering a new material/technology, or from a customer with an unusual request. In deciding whether or not to pursue an idea, it is necessary to consider the potential business value.

Any idea has to be **commercially viable**; it is no good coming up with an idea which sounds great but nobody wants to buy. The product should also have benefits for both consumer and manufacturer. Concept sketches, modelling and **CAD** are all used to communicate ideas and to arrive at a final design which defines, in detail, what the final product will be.

**best practice** – the best possible way of doing something

**genesis** – the origin of something

**commercially viable** – able to succeed in the commercial sector

**CAD** – computer-aided design

**revenue** – income

This is, however, not the end of the planning and design stage. Indeed, changing government legislation or currency fluctuation may mean the initial concept needs to be redeveloped or even scrapped. Increased costs brought about by, for example, currency fluctuation can be absorbed more readily by a large company who can manufacture products on a large scale to generate increased **revenue**. Small business enterprises do not, normally, have the available funds to take such risks.

The next stage is to turn the initial idea into a real business opportunity. You need to be clear about the investment of time and money needed to develop your product and the potential return for your investment.

## New products designed and developed by Mike Satur

Mike Satur has designed and developed some unique products for the MG: a roll hoop, a windbreak and a seat belt holder.

### Roll hoops

Roll hoops are essentially two roll bars, one behind the driver and one behind the passenger.

MG with roll hoops

The Boxer roll hoops were designed, developed and prototyped by Mike Satur in 2000. The way the system attaches to the car, the shape and the aesthetic appeal are unique to this design, and great care was taken not to copy any other designs that were then available.

The design incorporates high-strength steel tubing, as found in roll cages. The structural steel subframe attaches to the car's structure by utilising the existing mounting points on the car. This eliminates the need for any drilling, which could affect the **structural integrity** of the car.

The roll hoops are manufactured **in-house** to strict standards so that every one will fit and function as designed.

HEALTH & SAFETY

**Drilling holes in the bodywork of a car will affect the strength of the material, which could reduce the degree of safety protection.**

## Windbreak screen

The windbreak screen is constructed of **toughened glass**. Again, this product can be fitted without the need for drilling.

MG with windbreak screen

**structural integrity** – measure of the quality of construction and the ability of the structure to function as required

**in-house** – done or existing within an organisation

**toughened glass** – commonly known as safety glass

**ingenious** – clever, original and inventive

## Seat belt holder

The seat belt holder is an example of an **ingenious**, yet simple, idea. The original MG design meant the seat belt rested behind the seat, which made finding and retrieving the belt very difficult, due to limited space.

MG seat belt holder

The addition of the holder ensures the seat belt remains in position, enabling ease of use.

# Costs

Any company has two types of cost – **fixed costs** and **variable costs**.

Fixed costs stay the same no matter how many products are produced, and include:

- the rent on buildings
- the wages paid to workers
- advertising.

**FIG 4.10:** Company fixed costs against total units sold

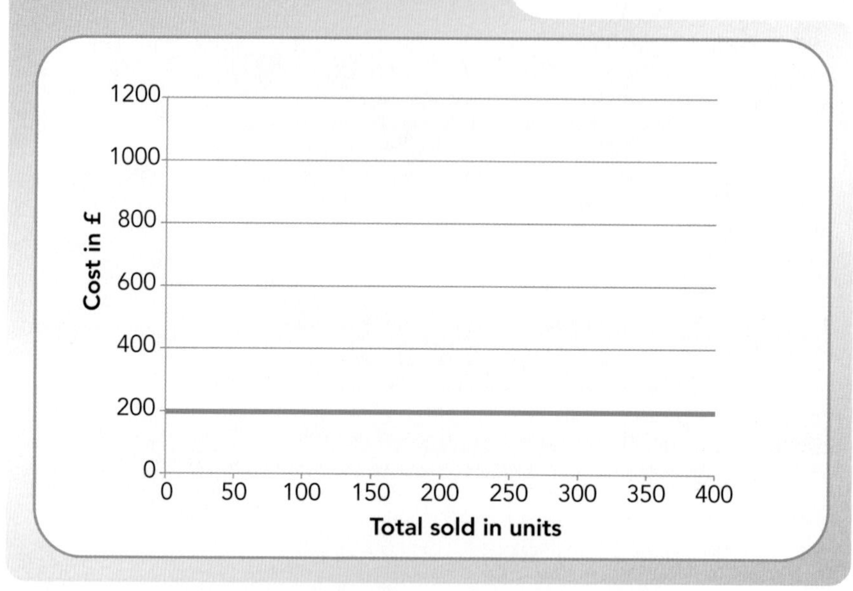


**MATHS & SCIENCE**
The horizontal line on the graph indicates a constant value, i.e. no change.

Fixed costs must be paid even if there is zero production.

**FIG 4.11:** Company variable costs against total units sold

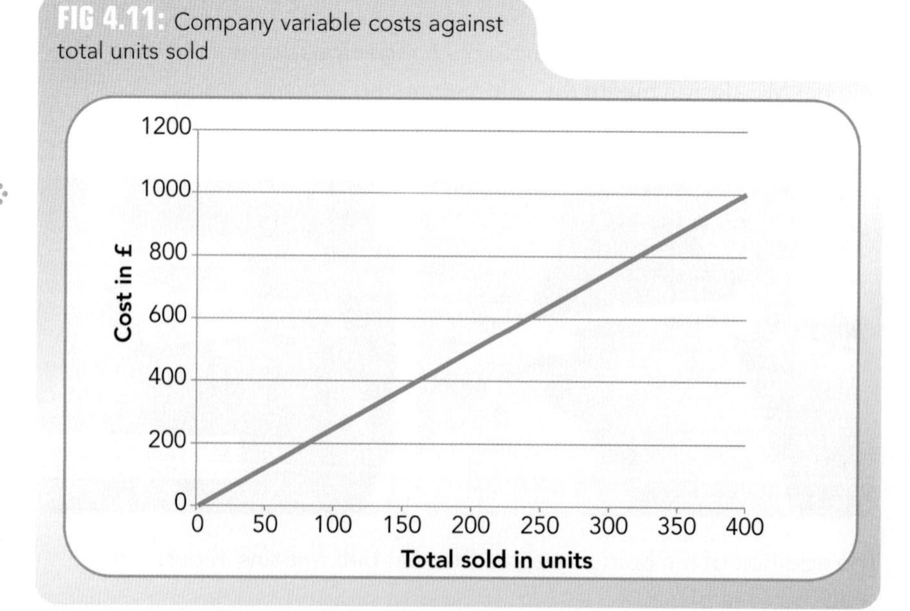


**MATHS & SCIENCE**
A linear graph is one in which all the plotted points lie on a straight line.

Variable costs go up and down according to how many products are produced. Variable costs therefore include:

- the costs of the materials to make the products
- transport costs.

Fixed costs are ones that do not change with the level of sales. Variable costs are ones that change in **proportion** to sales.

**proportion** – the number, amount or level of one thing when compared to another

**break-even** – to have no profit or loss at the end of a business activity

FIG 4.12: Total costs against total units sold

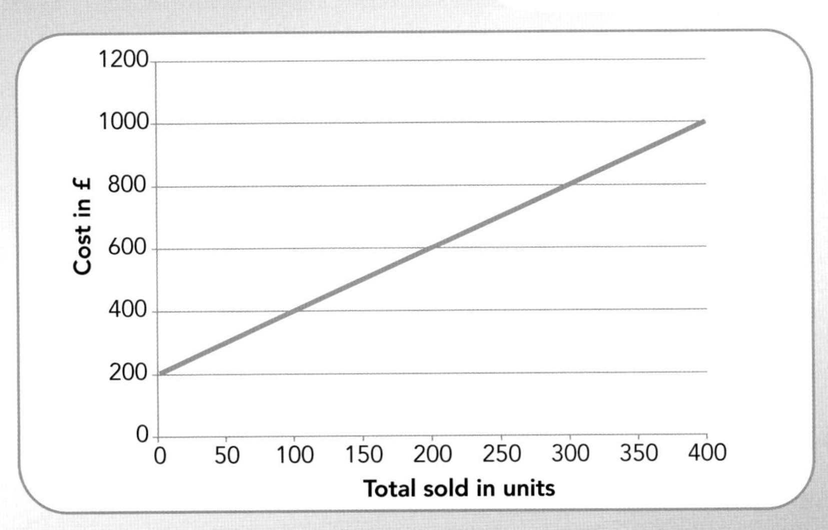


FIG 4.13: Total costs and sales income against total units sold

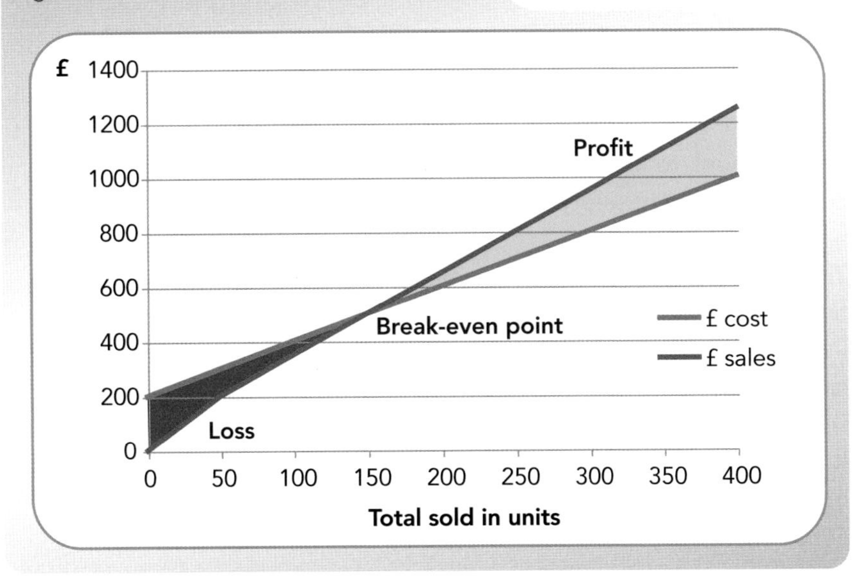


**MATHS & SCIENCE**

**On a graph, the break-even point is found at the intersection of the 'sales income' and 'total cost' lines.**

By putting the fixed and variable costs together, a business can calculate what is called a **break-even** point. This is where the amount of money needed to manufacture the product is covered by the money generated through sales of the product.

Engineering companies, like other companies, are always looking for ways to cut costs to ensure profitability.

The most expensive fixed costs are usually employee wages. That is one of the reasons why, for mass-produced goods, robots have replaced people. However, robotic systems are expensive to buy and install and, in the case of Mike Satur, where the production is on a smaller scale, a robot is not the answer.

An engineering company has another very difficult decision to make: do they make the product themselves or buy it in from elsewhere. Occasionally, this decision is straightforward as the company may not have the necessary tools, equipment or expertise to make the product themselves.

When a company has to purchase products from other suppliers it has to be very careful about how many it orders at a time. The more it orders the cheaper they will be, but if the product then stays on a shelf in a storeroom and fails to sell, the company has not generated any income to cover its purchase, so it is operating at a loss.

## Production and testing

Before manufacturing a product, a prototype will be made to test the design and to check that it meets the customer's requirements. Further modifications and changes may be necessary up to the point of production.

The final product needs to be thoroughly tested before being launched on the market. In the case of Mike Satur, products must undergo rigorous safety testing. It may also be necessary, even at this late stage of product development, to further modify the product to meet both current and, possibly, imminent safety legislation.

The launch of the product on the market may require extensive publicity, and this is probably one of the most difficult and expensive stages. It includes developing the market and deciding on the most profitable **sales channels**.

**sales channel** – a way of selling a company's products directly or via distributors

It is vital to keep reviewing and modifying the product so that, as sales of one product decline, new or modified products are introduced. Mike Satur has, in 2007, modified the roll hoop design to generate fresh interest in the product. This constant reviewing and upgrading of products is essential for the survival of a small business.

# The career path

## PERSONAL PROFILE: Mike Satur

**What qualifications do I need to become a civil engineer?**

You need to start with passes at GCSE (or equivalent), to include maths and a science subject. There are many routes you can take from there to begin your journey towards a career in the civil engineering industry.

For example, you could take A/AS levels or Highers followed by a MEng or BEng degree course at university.

Other options include the BTEC National Diploma/Certificate in Construction, or a technical apprenticeship or similar work-based programme.

### Why did you decide to start your own business?

I had been in the civil engineering industry since the 1960s. I had got to the stage where I felt that there were no more doors to be opened; there were no further challenges for me.

By the 1980s my interest in the motoring industry was stronger than my interest in the civil engineering industry.

I decided to take the plunge and went into the motor industry on a full-time basis, and I have never looked back.

### Why the MG car?

It is well known that they are not the most reliable of cars, but they are fun to drive. That is the important factor: the fun element.

### Why do customers keep returning?

If you run your own business, you need to be interested in what you do. I am an enthusiast and that is what attracts people to this business. I have the same obsession as the customer.

**NUTS & BOLTS**

**To find out more about an apprenticeship in the construction industry go to www.citb.co.uk/traininglearning/nationalconstructioncollege/courses/apprenticeships**

**NUTS & BOLTS**

**To find out more about a career in the automotive industry go to www.prospects.ac.uk and www.autoindustry.co.uk**

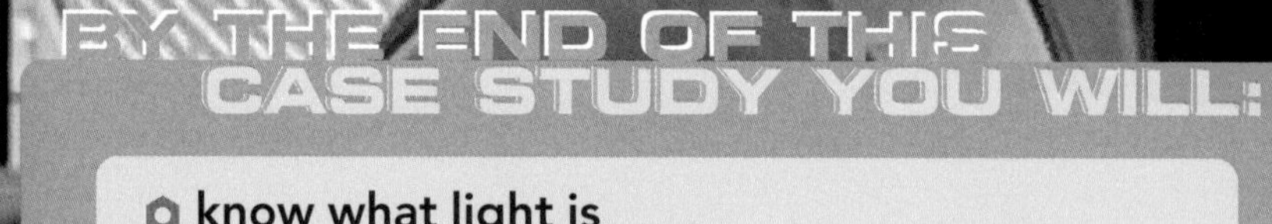


- know what light is
- have a basic understanding of reflection and refraction of light
- have become familiar with some lighting system terminology
- have an appreciation of the design elements involved in a lighting system
- be able to recognise how light is used to influence the mood and function of a lit space
- have some ideas as to how you can become a lighting designer.

# BUILDING SERVICES – LIGHTING DESIGN: TROUP BYWATERS + ANDERS

## Case study overview

This case study introduces you to the building services sector and, specifically, lighting system design.

Over the following pages you will become familiar with a real company that designs and engineers lighting systems for a range of clients.

After considering the basic principles of light, you will follow the company's procedures through consideration of the design of a lighting system for two contrasting spaces. This will help you to understand the importance of light in real-life situations and enable you to evaluate the effectiveness of the design solution.

Finally, you will be introduced to a lighting designer from the company who will describe the route she took to her current position.

# The sector

## The building services sector

Building services engineering plays an important part in all our lives – almost every building contains some form of electrical power, heating, lighting, air conditioning, ventilation, refrigeration and plumbing.

The building services sector represents the **electrotechnical**, heating, ventilation, air conditioning and plumbing industries. The UK's building services sector carries out £20 billion of work each year – 3 per cent of **GNP**, and employs almost 600,000 individuals in over 50,000 businesses.

At any one time, the building services sector is training 18,000 apprentices.

**electrotechnical** – electrical, electronic and communication technologies
**GNP** – gross national product: total value of goods produced and services provided by a country during one year
**blueprint** – a design plan or other technical drawing

## What is lighting design?

Lighting design engineers design and engineer lighting systems that provide illumination in places such as offices, sports venues, hospitals and retail outlets. They may also focus on using light to enhance, decorate and emphasise works of art, artistic performances (such as theatre shows and music concerts), buildings and public monuments.

Lighting engineers must understand the design, installation, maintenance and operation of lighting systems. This requires knowledge of electrical wiring and engineering, and the ability to understand **blueprints**.

NUTS & BOLTS

To find out more about the building services sector go to www.modbs.co.uk

# The company

## Background

The Troup Bywaters + Anders (TB+A) partnership has been delivering building services engineering consultancy to a growing range of market sectors since 1958.

They are one of the country's leading building services engineering consultancies, with a team of multidisciplined designers working across public, private and joint sectors, including defence, education, retail and transport.

Throughout their work they continue to reappraise construction methods and apply **innovative** approaches to ensure that the buildings they create enrich the lives of the people who use them, whilst having minimal impact on the environment.

The TB+A working environment and practice **culture** are developed to empower their staff to achieve their personal goals, alongside delivering the **optimum** solutions for their clients. As a consequence, TB+A staff remain with the company for much longer than the average in the building services industry.

### Light and lighting solutions

One of the services provided by TB+A is that of lighting design. This encompasses not only design but regulatory advice, surveys and reports, production of detailed design drawings and specifications, and daylight modelling. Clients employing their services have included the BBC, the jewellers Asprey and Garrard, and Virgin.

**innovative** – advanced and original
**culture** – the attitudes and behaviour of a particular group of people
**optimum** – best, most likely to bring success
**cumulative total** – a running total

## Modern-day concerns

Artificial light accounts for almost one-fifth of the world's electricity consumption, substantially more than the output of all the nuclear power stations in the world. It generates around 1.9 billion tons of carbon a year, equivalent to nearly three-quarters of the carbon coming from the exhaust of all cars and light vehicles in the world.

The UK government alone has 50,000 buildings, with a combined annual energy bill of almost £200 million. The fact is that most of these buildings are using inefficient lighting systems.

The low-energy light bulb and other efficient lighting systems could prevent a **cumulative total** of 16 billion tons of carbon from being added

**HEALTH & SAFETY**

**Carbon dioxide emissions contribute to climate change, which could have devastating effects on the quality of life for the future.**

**Kyoto Protocol** – sets binding greenhouse gas emissions targets for countries that signed and ratified the agreement

NUTS & BOLTS

**To find out more about TB+A go to www.tbanda.co.uk**

to the world's atmosphere over the next 25 years, according to a report by the International Energy Agency.

Moreover, a high-quality energy-efficient lighting system can reduce operating costs – a significant advantage to industry.

Since the **Kyoto Protocol**, the lighting industry has been responding to the call for energy-efficient lighting to reduce emissions of carbon dioxide. Moreover, the UK government has introduced regulations designed to improve the overall energy efficiency of lighting installations.

It is the lighting design engineer's greatest challenge: providing conditions for users to carry out their tasks safely, comfortably and with high productivity whilst ensuring the energy efficiency of lighting and lighting installations.

# The product

## Introduction

Lighting is one of the most difficult concepts in building services engineering. Light is all around us and is responsible for everything we see, but we cannot touch it or feel it. The way that light is used can change the way that we see everything; it can also affect our mood and the feel within a building.

When most people are asked what they think lighting design is about, their immediate thoughts are pop concerts and the theatre, where light, shade, colour and tone are used to create dramatic effects. Indeed, it is the job of lighting engineers to create these effects, but lighting is not restricted to pop concerts.

Lighting is essential in every building, and lighting engineers have to consider not only the balance between light, shade and colour but the specific lighting requirements for a working environment, i.e. what tasks are to be carried out in the space. They also have to consider the fixtures and fittings for the light sources, how to provide the power source and, critically, they must allow for easy maintenance.

To an untrained eye, unlike a pop concert where you enjoy the dazzling effects of light, in a building it is only bad lighting that is noticed, or, more commonly, people are only aware of lighting when it is not working.

# What is light?

If you throw a stone into a pond, the splash will cause ripples in the form of waves that **radiate** from the place where the stone hit the water. The water itself does not move out from the entry point of the stone: waves are not water, they are energy moving through the water.

All waves are travelling energy, and they are usually moving through some **medium**, such as water. The size of a wave is measured by its **wavelength**. This is the distance between any two corresponding points on **successive** waves.

**radiate** – spread out in all directions from a central point
**medium** – a substance
**successive** – following one another
**vacuum** – a space entirely devoid of matter
**spectrum** – the entire range of wavelengths of electromagnetic radiation

FIG 5.1: A wave form

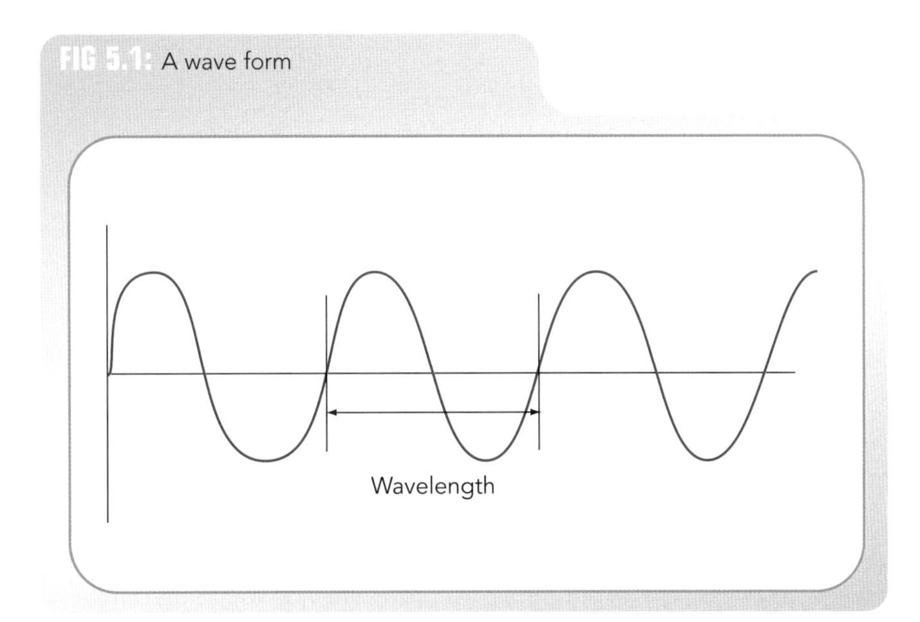


Understanding wavelengths is a vital part of understanding light. Light is made up of waves of energy in the form of electric and magnetic fields. Due to the electric and magnetic fields, light is also referred to as **electromagnetic radiation**. Unlike other waves, however, light can travel through a **vacuum**. It radiates equally in all directions, but as it travels away from its source it gets less intense.

The wavelengths of the light we can see range from 400 to 700 billionths of a metre. In fact, light is one small part of the **spectrum**. The full range of wavelengths extends from one billionth of a metre (gamma rays) to centimetres and metres (radio waves).

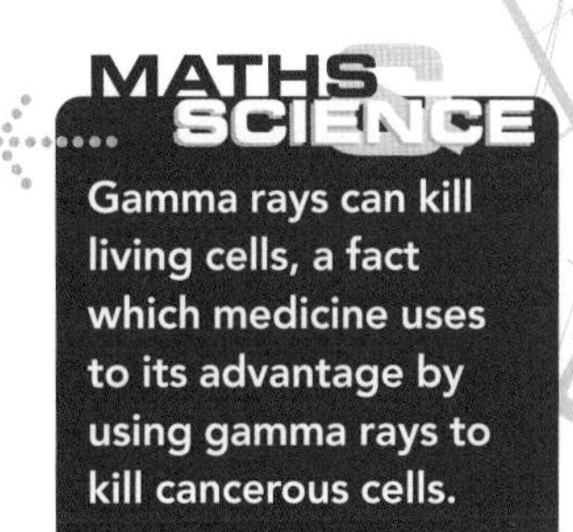
MATHS SCIENCE
**Gamma rays can kill living cells, a fact which medicine uses to its advantage by using gamma rays to kill cancerous cells.**

Light waves also come in many different **frequencies.** The frequency is the number of waves that pass a point in space during any time interval. As the frequency increases, the wavelength decreases.

Lower frequencies result in red light, while higher frequencies result in blue light. Red light is at one extreme end of the visible spectrum, while blue or violet light is at the other.

**MATHS & SCIENCE**

Nothing can travel faster than electromagnetic waves. The speed of sound in air is approximately 300 m/s, so light is almost 1 million times as fast.

Light travels at different speeds. Light waves move through a vacuum at their maximum speed (300,000 km/s). They slow down when they travel inside substances, such as air, water and glass.

## The colour of light

The name given to the light we can see with the human eye is visible light. When you look at the visible light of the sun, it appears to be colourless, which we call **white light**. Though we can see this light, white is not considered to be part of the visible spectrum. This is because white light is not the light of a single colour or frequency. In fact, it is made up of many colour frequencies.

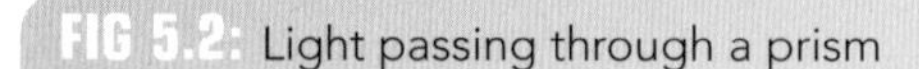

FIG 5.2: Light passing through a prism

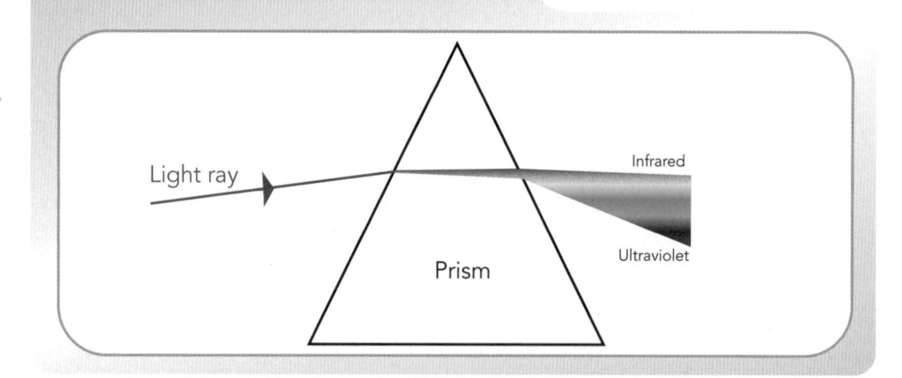


**MATHS & SCIENCE**

Red light has the longest wavelength and is refracted the least. Violet light has the shortest wavelength and is refracted the most.

In 1666, Sir Isaac Newton conducted experiments with a glass prism and sunlight. By passing sunlight through a prism, Newton found that white light separates into the same progression of colours found in the natural rainbow. This would not have happened unless white light were a mixture of all of the colours of the visible spectrum.

**MATHS & SCIENCE**

Newton discovered the nature of light and colour, developed the laws of motion, and much more!

Using a second prism, Newton returned the constituent colours back into white light. This proved conclusively that each colour in the spectrum was pure – composed of a single, unique wavelength – and that coloured light mixed together gives white light.

If you look at the Wyndeham Westway case study, you will learn about two types of colours: additive and subtractive colours.

### Colour by addition

By adding various combinations of red, green and blue light, you can make all the colours of the visible spectrum. This is how TV and computer monitors, known as RGB (Red Green Blue) monitors, produce colours.

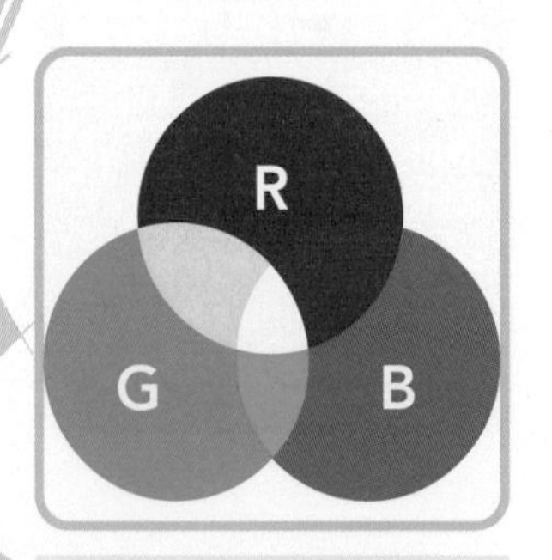


Colour by addition

### Colours by subtraction

Another way to make colours is to absorb some of the frequencies of light, and thus remove (subtract) them from the white light combination.

You do not see the absorbed colours; you only see the colours that come bouncing back to your eye. This is what happens with paints; the paint **molecules** absorb specific frequencies and bounce back, or reflect, other frequencies to your eye.

Nature does this; the leaves of green plants contain a pigment called chlorophyll. Chlorophyll absorbs the blue and red colours of the spectrum and reflects the green. Leaves therefore appear green.

**When light enters the eye, it first passes through the cornea. Ultimately, it reaches the retina, which is the light-sensing structure of the eye.**

**molecule** – the smallest unit of a chemical compound

**emitting** – sending out

**incidence** – the intersection of a beam of light with a surface

# What happens when a light wave hits an object?

When light hits an object:

- the waves can be **absorbed** by the object
- the waves can be **reflected** or scattered off the object in different ways
- the waves can be **refracted** through the object
- the waves can **pass through** the object with no effect.

More than one of these possibilities can also happen at once.

## Absorption

As noted above, the colour of an object depends on the wavelengths of the colours reflected from the object. When a black object is illuminated by white light, all the wavelengths are absorbed – none are reflected – and this is why the object appears black.

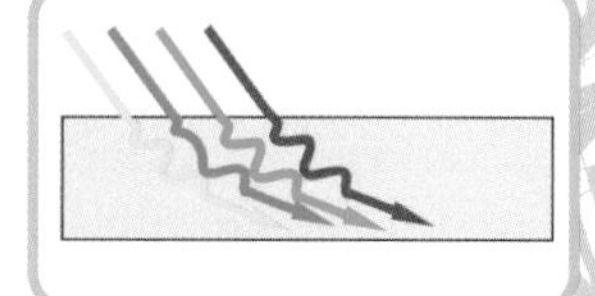

Absorption

When light is absorbed by a black object, the energy carried by the light does not just disappear. In fact, it raises the energy of the object doing the absorbing, which then releases the absorbed energy by **emitting** longer-wavelength, lower-energy infrared radiation – heat.

The darker the object, the better it is at emitting heat. This is because it is a better absorber of light. So, dark materials get hotter in the sun because of this absorption effect. Solar panels are an excellent example of how engineers can use light absorption to good effect.

MATHS SCIENCE

**Energy cannot be created or destroyed. It may be transformed from one form to another, but the total amount of energy never changes.**

## Reflection

Reflection is the change in direction of a wave when it bounces off a boundary.

A reflected light wave always comes off the surface of a material at an angle equal to the angle at which the incoming wave hit the surface. In physics, this is called the **law of reflection**. The law of reflection states that 'the angle of **incidence** equals the angle of reflection'.

Reflection

**MATHS & SCIENCE**

Electromagnetic waves are sometimes called rays.

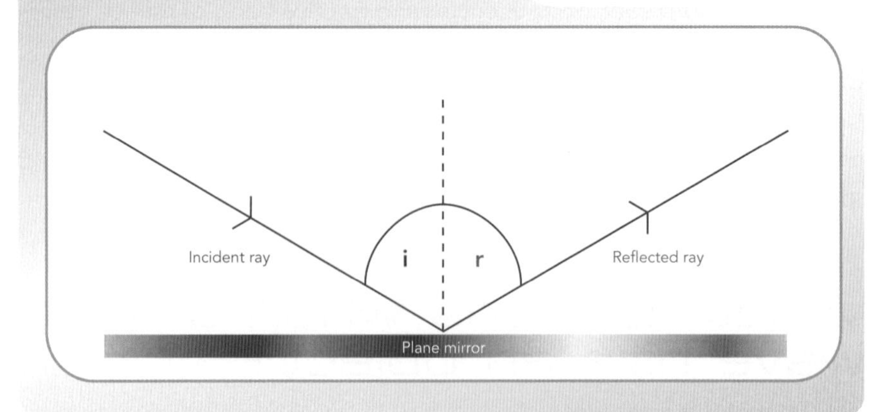


FIG 5.3: Angle of incidence equals angle of reflection

After reflection, a wave has the same speed, frequency and wavelength; it is only the direction of the wave that has changed.

**velocity** – the speed of something in a given direction

A mirror is a perfect example of a reflective surface. The light is reflected with the same frequency as it hits the surface. All the colours remain as before, with no absorption.

**Scattering**, or **diffuse reflection**, is the same as regular reflection but the reflection is off a rough surface. Incoming light waves get reflected at all sorts of angles because the surface is rough or uneven.

Scattering or diffuse reflection

This is the most common type of reflection, as most surfaces are irregular when considered on a scale comparable to that of the wavelength of light.

## Refraction

Refraction is the bending of a wave as a result of its **velocity** changing when it moves from one medium into another.

**MATHS & SCIENCE**

Half-submerge a straight stick in water. Does the stick appear bent at the point it enters the water? This optical effect is due to refraction.

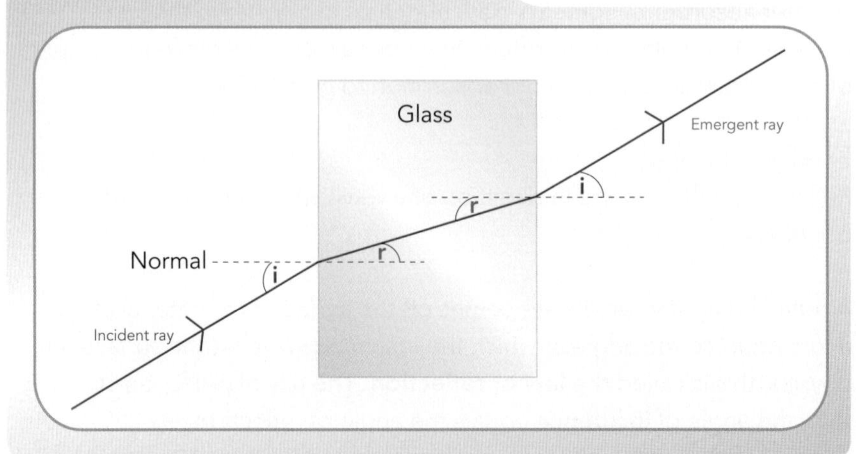


FIG 5.4: Angle of incidence is not equal to angle of reflection

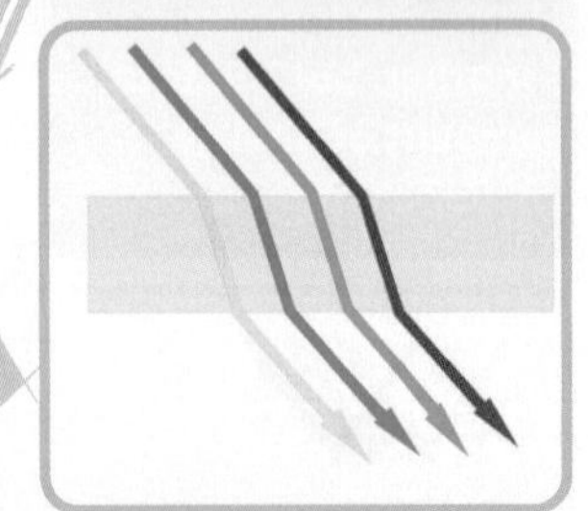

Refraction

After refraction, the wave has the same frequency but a different speed, wavelength and direction.

When light waves move into a new medium, they are refracted. The direction in which they are refracted depends on whether they move into a denser or less dense medium and are consequently slowed down or speeded up.

If the second medium is denser, the wave slows down and is refracted towards normal – an imaginary straight line that is perpendicular to the surface of the object.

If the second medium is less dense, the wave speeds up and is refracted away from normal.

Materials can be classified by their **index of refraction**. Diamonds have a high index of refraction as they slow down light to a greater degree than many other materials.

It is important to note that light is only refracted when it hits a boundary at an angle. If light goes straight down into an object it will continue to move straight down.

**MATHS & SCIENCE**

**Rainbows form when light is refracted through tiny drops of rain present in the air after a shower. Each drop acts like a prism, dispersing light into the colours of the visible light spectrum.**

# How do we measure light?

**Luminance** is a measure of how much light a bulb produces, measured at the bulb. Luminous intensity is measured in **candela** (cd). Originally, luminous intensity was measured in terms of units called candles. This term came about from the fact that one candle represented approximately the amount of visible **radiation** emitted by a single candle flame. This was an inexact specification for engineers because different types of candles give out different amounts of light when they burn. Late in the twentieth century, the current definition and terminology were adopted.

Today, a candela is defined for radiation at a single frequency.
To be precise:

> a candela is formally defined as the luminous intensity, in a given direction, of a source that emits **monochromatic** radiation of frequency 540 x $10^{12}$ hertz and that has a radiant intensity in that direction of 1/683 watt per **steradian**.

The watt (a unit of power) was not initially tied to the candela because of the eye's differences in seeing various wavelengths of light. Since the human eye is less sensitive to blue and red light, more wattage is needed to produce the same result in the brain as with yellow or green light. This

**index of refraction** – a measure of how much the speed of light is reduced inside a medium

**radiation** – electromagnetic waves

**monochromatic** – a single wavelength or frequency

**steradian** – the SI unit of solid angle

is why the definition of candela is defined as a very specific hertz, unachievable in actual common lighting.

Since the precise definition is so unwieldy, it is not uncommon in standard usage to see the candela referred to as 'roughly' the amount of light generated by a single candle. 120 candela is roughly the light emitted by a 100 W light bulb.

**MATHS & SCIENCE**
**SI is the international system of units of measurement.**

**perceive** – interpret or regard something in a particular way

The **lumen** (lm) is the SI unit of **luminous flux**, a measure of the **perceived** power of light. A lumen can be thought of as a measure of the total 'amount' of visible light emitted.

**Illuminance** is what results from the use of light. It is the intensity or degree to which something is illuminated, and is not therefore the amount of light produced by the light source. When you are measuring illuminance you are looking at how much of the energy released by a light source makes it to a given object, i.e. the light level on a particular surface.

The **lux** (lx) is the SI unit of illuminance and is the measurement of actual light available at a given distance. A lux equals one lumen per square metre of illuminated surface area.

The difference between the lux and the lumen is that the lux takes into account the area over which the luminous flux is spread. 1000 lumens, concentrated into an area of one square metre, lights up that square metre with an illuminance of 1000 lux. The same 1000 lumens spread out over ten square metres produces a dimmer illuminance of only 100 lux.

## Exploring the common light bulb

Can you imagine what it would be like to live without electric light bulbs? After sunset you would have to use candles and oil lamps to light your house.

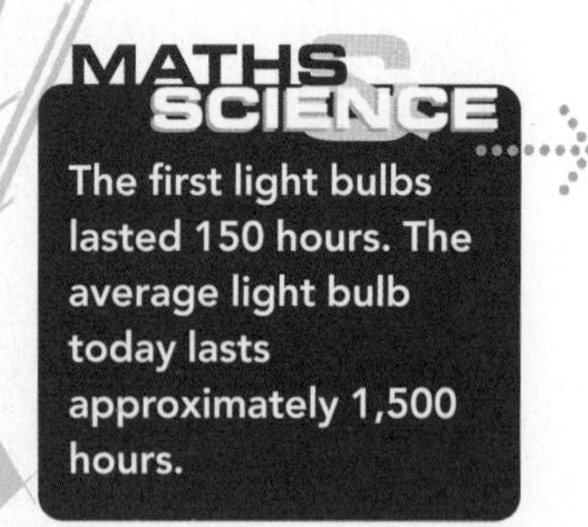


The first light bulbs were designed by the mid 1800s and, interestingly, modern light bulbs have not changed very much from the first designs.

### The bulb

Light bulbs consist of a metal base with a screw thread or push fitting contact. The metal contacts at the base of the bulb are connected to two wires that go to the middle of the bulb and hold the filament in place. This is the part of the bulb that creates the light. It consists of a long length of very thin tungsten wire which is tightly wound into a double coil.

FIG 5.5: Parts of a light bulb

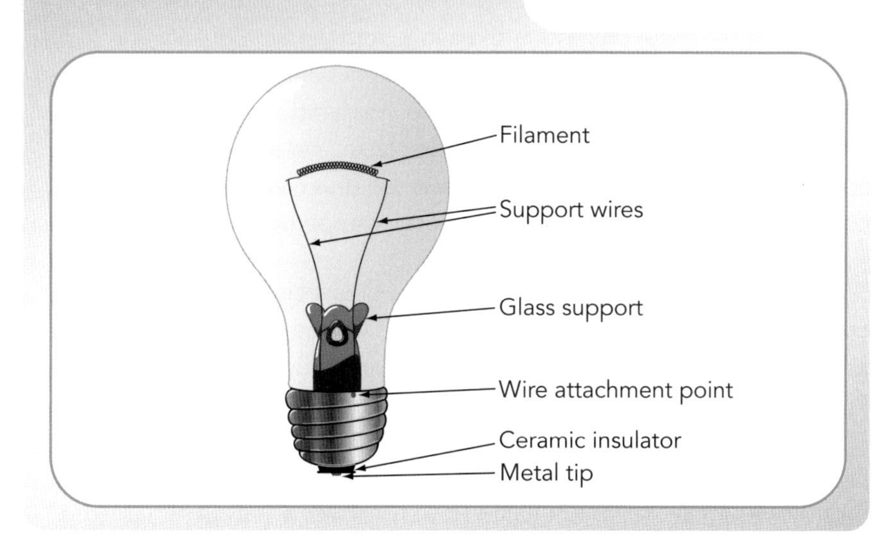


The bulb itself is made of glass, which protects the filament from oxygen in the atmosphere but also keeps in an **inert gas**, usually argon.

### How does a light bulb work?

When electricity is applied to the bulb, it passes through the contacts to reach the connecting filament, which gets very hot (over 2,000 degrees Celsius). This causes it to get white-hot and thus glow, emitting a good deal of visible light.

Bulbs also emit lots of infrared light, which cannot be seen. Infrared light is pure heat, and is the reason why bulbs get so hot. In fact, because of the way light bulbs work, most of the radiation is emitted as heat rather than visible light – only about 10 per cent of the light produced is in the visible spectrum. For this reason, light bulbs are not very efficient generators of light by today's standards.

**MATHS & SCIENCE**

**Tungsten is chosen as the filament as it has an abnormally high melting temperature.**

**MATHS & SCIENCE**

**The inert gas is used to protect the filament from burning out.**

**inert gas** – any one of six gases that does not react with other substances under ordinary conditions

## Electricity

### Conductors and insulators

The lighting in a building is dependent on the electricity supply to the building. All electrical systems have to comply with building regulations.

The electrical energy flows through a conductor. Metals make good conductors as they have low resistance. In fact, copper is one of the most often used conductors of electricity as it has a very low electrical resistance.

The electrical conductors need to be insulated from other parts of the building. Wires and cables therefore consist of an electrical conductor surrounded by an insulator.

**MATHS & SCIENCE**

**A conductor is a material containing a large number of charges which are free to move. It can therefore conduct electricity.**

Most of the wires and cables used in buildings are marked by a standard system to indicate the safe voltage they can carry, the number and size of the electrical conductors within the cable, and the type of insulation. The purpose of the insulation is not just to prevent contact with other metal objects; it also protects against moisture, heat and corrosion. Materials that are good insulators include rubber, plastic, ceramics and glass. In some buildings, armoured cable is used too (this consists of two or more conductors protected by a flexible, metal wrapping).

To provide further protection, wires and cables are often enclosed in a **conduit**. A metal conduit is often used as it can be permanently **grounded**. Metal is also often used to provide fireproofing for the cable. Plastic conduit is more commonly used for underground wiring.

**conduit** – a tube for protecting electric wiring
**ground** – to earth an electrical appliance
**electromotive force** – a difference in potential which causes current to flow in a circuit

Deciding the best route for the wiring within a building can be a difficult task. The building will also have to carry mechanical and plumbing systems, and it is important to avoid conflicting pathways. In order to maintain the system, access to the fittings is also a key consideration.

## Units of electrical energy

Electrical energy flows through a conductor because of a difference in electrical charge between two points in a circuit. Electrical energy is measured using the following methods:

- **Volt** (V) is the SI unit of electric potential or **electromotive force**. A potential of 1 V appears across a resistance of one ohm when a current of one ampere flows through that resistance.
- **Ohm** (Ω) is the standard unit of electrical resistance. One ohm is the equivalent of a volt per ampere (V/A).
- **Ampere** (A), often shortened to amp, is a unit of measure of the rate of electron flow or current in an electrical conductor.
- **Watt** (W) is the SI unit of power (energy per unit time), equivalent to one joule per second. The watt is used to measure power or the rate of doing work.

A human climbing a flight of stairs is doing work at the rate of about 200 watts. A first-class athlete can work at up to approximately 500 watts for 30 minutes. A car engine produces 25,000 watts of mechanical energy while cruising. A typical household light bulb uses 40 to 100 watts.

## Supply

The power is supplied to the building by an electricity supply company. The connection to the company can be overhead or underground. The advantage of an overhead supply is that it is easily accessible for maintenance.

To ensure a minimal loss of power, the power supply is often transmitted at very high voltages. Because of this, most medium and large buildings need a power transformer. Quite often these transformers are outside the building as they need to be well ventilated.

Having been **stepped down** to a usable voltage the power will flow in to a main switchboard, which is usually located close to the transformer to minimise any drop in voltage and to reduce the amount of wiring needed.

**step down** – to reduce the amount, supply or rate of something

Once the electrical power requirements for various areas of the building have been established, wiring circuits need to be designed to distribute the power to where it is needed. Separate wiring systems are required for sound systems, telephone systems, lighting and security and fire alarms.

Lighting fixtures, wall switches and wall sockets are usually the only parts of an electrical system that are visible within a building.

# Lighting

Lighting types are classified by intended use as general, localised or task lighting, depending largely on the distribution of the light produced by the fixture.

Task lighting is mainly functional and is usually the most concentrated, for purposes such as reading or the inspection of materials.

Accent lighting is mainly decorative, intended to highlight pictures, plants, or other elements of interior design or landscaping.

## Methods of lighting

**Downlighting** (downward illumination) is possibly the most conventional type of lighting used in buildings. It provides direct and energy-efficient light to a specific area. It is energy efficient as it focuses the light directly to where it is required.

Downlighting and backlighting

Direct lighting helps us to see shape, form and texture by casting shadows and variations in brightness. In buildings, downlights are often recessed into the ceiling.

The light produced from downlights is usually a three-dimensional cone of light.

Although downlighting is easy to design, it has dramatic problems with glare. Glare occurs when bright light sources interfere with the viewing of objects or surfaces that are less bright.

**HEALTH & SAFETY**

**The contrast between very bright and less bright light may be uncomfortable or disabling, both of which are undesirable.**

Uplighting

**Uplighting** is the general term for lighting from below.

This technique is achieved by placing a lamp at the base of the object to be lit. Light is cast upward.

Because uplighting is not a natural way of lighting, it tends to draw our attention to the object being uplighted. Uplighting is often used to make a room appear more spacious. Light is directed up at the ceiling which acts as a reflector and, as such, bounces the light back down – giving the appearance of 'light' in the room.

Uplighting is commonly used in lighting applications that require minimal glare and uniform general levels of illuminance.

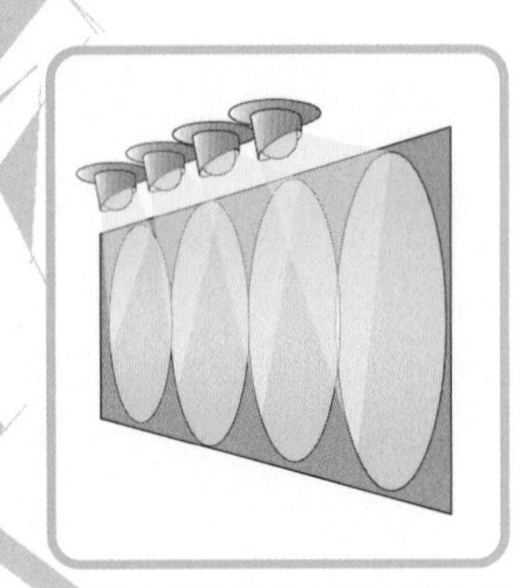
Wall washing

**Wall washing** is often used to emphasise vertical surfaces in a room.

Wall washing is also used to emphasise the wall texture, pictures or features. Special fixtures are used to flood a wall with light.

Wall washing is perfect for retail stores, classrooms and other spaces with wall displays.

**Feature lighting** is used to highlight features in a room whilst hiding less significant aspects of the room. Feature lighting usually works best when the light source itself remains hidden.

## Light fittings

Light fittings, or **luminaires** as they are called by the industry, are the equipment that physically supports the lamp and provides its safe connection to the electricity supply. Light fittings also provide protection for the lamp, particularly in hazardous areas and areas where broken glass would be a particular problem, e.g. food processing.

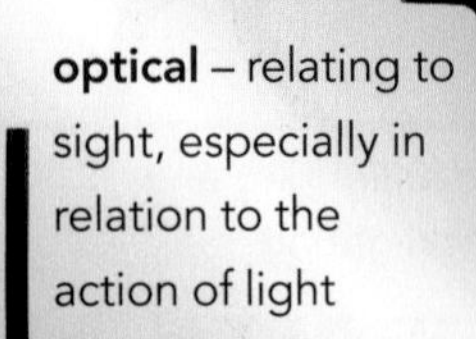
**optical** – relating to sight, especially in relation to the action of light

Luminaires also provides the **optical** control that ensures the light is directed to where it is required, as well as obstructing it from those areas where it is not needed. This involves the use of reflectors, refractors and/or diffusers.

The optical elements of the luminaire absorb light. This means that not all the light from a lamp will emerge from the luminaire, as some will be used in the process of redirection to create the light output pattern required.

Look at the types of fixtures used in the illustrations in Figure 5.6 and how the light is affected above and below the horizontal plane.

Luminaires come in a wide variety of styles for various functions. Some are very plain and functional, whilst some are pieces of art in themselves.

**FIG 5.6:** Pattern of light output produced by different fixtures

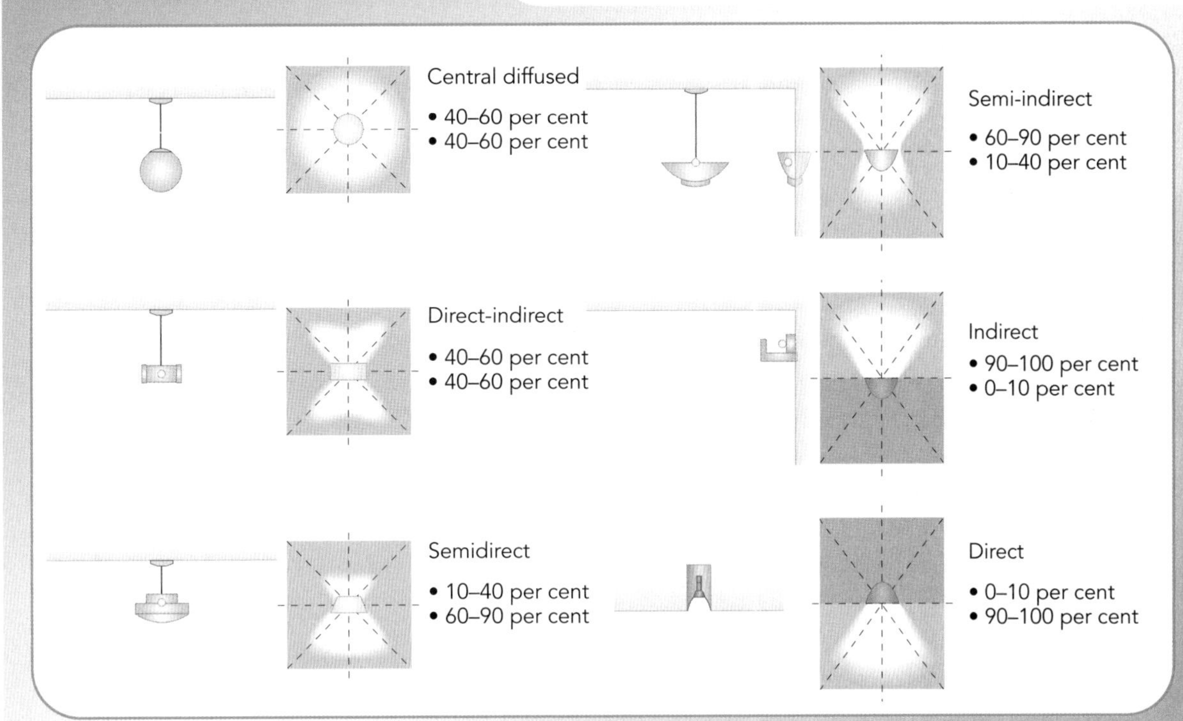


Nearly any material can be used, as long as it can tolerate the heat and is in keeping with safety codes.

# The process

## Design process

Light plays an essential role in our ability to perceive the world around us; the lighting system plays a critical role in how we perceive a space, and can influence how we act in the space. Lighting can affect performance, mood, morale, safety, security and decisions.

The first step in producing the right lighting design is to ask what the space is used for. The lighting designer can then determine quantity of light, colour quality, brightness and direction. Consideration also needs to be given to any supplementary lighting which may be required for safe circulation and to create a building that has a pleasant appearance.

When designing an interior lighting system the designer also needs to consider the types of materials within the space to be lit, and whether

**contour charts** – charts marked with lines connecting points of equal light

**mock-up** – a model or replica used for demonstrational purposes

HEALTH & SAFETY

The Society of Light and Lighting provides recommendations for task illuminance for a range of different situations.

permanent features, such as windows, doors, wiring and ductwork, will dictate certain light layouts.

Once the designer understands what limitations the building puts on the designs, he or she will need to consider the recommended task illuminances to use for each task area. The effects of the décor and surface reflectance in each area will need to be taken into account. Unless these are defined in the brief, the designer should confirm to the client the lighting levels being used for the design and any assumptions being made about surface reflectance or maintenance cycles.

Computer modelling is used by the designer to simulate the effects of different types of light. Using lighting design software, each fixture has its location entered, and the reflectance of walls, ceilings and floors can be entered. The computer programme can produce a set of **contour charts** overlaid on the project floor plan, showing the light level to be expected at working height. This enables the designer to check that task areas have sufficient lighting, within regulatory guidelines, for their purposes.

It may also be necessary at this stage to produce **mock-ups** to allow the architect or client to see the effects of the lighting design.

## Putting light to work

**integrate** – combine

The lighting design engineer must work closely with the architect and client to **integrate** the lighting with the technical needs of the specific project, and the style of lighting or lit effect that has been visualised.

HEALTH & SAFETY

Regulations require that information is given about the building services so that the building can be operated in such a manner as to use no more fuel and power than is reasonable.

Correct implementation of the lighting system is a critical element. Consider the construction of a new building. A building is a complex product; its design and construction require the involvement of various disciplines and trades. Several designers and contractors are working on the same project simultaneously. Also, several subsystems, such as building services, need to be integrated for optimum performance of the building. The coordination of these many participants and subsystems is vital for a good overall result.

### Design 1: Office reception area

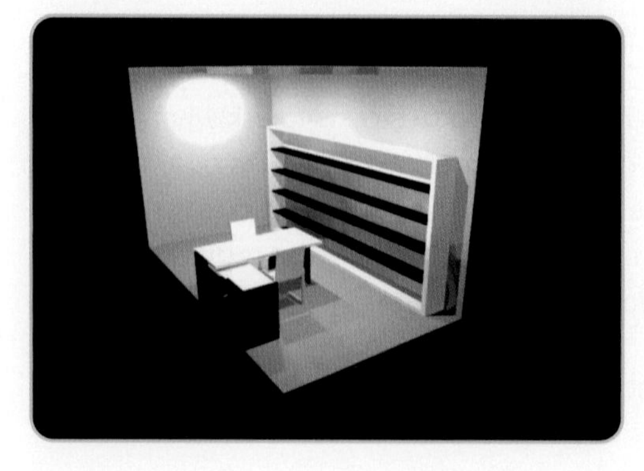

A typical reception area

The reception area is the first impression that customers have of a company. Space and lighting can play a major role in setting the correct atmosphere and reflecting the image of the company.

The reception area may be used for a variety of different tasks, so there must be adequate lighting provided for each task. The reception may include a waiting area as well as an office with a receptionist or secretary, so the engineer may need to plan for a variety of lighting in the same room, rather than uniform lighting across the whole room. This is in contrast to a normal office, where light should be distributed relatively uniformly, avoiding '**hot spots**', shadows or sharp patterns of light and dark.

**hot spot** – a small area of greater light

Reception area lighting needs to follow certain guidelines. The Chartered Institution of Building Services Engineers (CIBSE) recommends that reception areas should have 300 lux to allow the receptionist to work efficiently. Other areas, such as seating areas, can have less light, usually 200 lux, whereas corridors may only need 100 lux.

**300 lux will generally consume half the energy of an installation that provides 600 lux.**

Architect's model

The picture above shows the original design by the architect. The picture below shows the finished design.

Actual lighting system installed in office reception area

You will notice that the 'brightness' in the architect design is replaced by a 'subtle' lighting effect.

It is worth noting at this stage that when developing the original design idea, the architect is not taking into account design constraints such as wiring and ductwork. It may also be the case that the architect may not understand the effects of lighting and the necessary consideration that must be given to the reflection and absorption of light. It is the lighting design engineer's task to formulate a design which encompasses the architect's vision whilst ensuring that the lighting system can, in real life, be implemented and maintained with minimal difficulty.

## Design 2: Retail store

Retail today is extremely competitive; most stores have tight **margins** and are constantly looking for ways to attract consumers.

**margin** – profit margin: the amount by which income from sales exceeds costs
**LED** – light-emitting diode: a device which produces light

For retailers, lighting is an important means of portraying their image and company brand. For example, an upmarket store may require a sophisticated level of low-lights, whereas a large supermarket or warehouse need a more simple and efficient lighting solution.

Whatever the type of retail outlet, lighting can be used to attract the customer into the store, guide the customer around the store and highlight the merchandise on display. Results of studies have found that improving the lighting in retail outlets can significantly improve sales figures.

HEALTH & SAFETY

**High light levels can lead to extra energy being required for cooling. The high heat gains can result in many shops being uncomfortably hot, even in cool weather.**

Retail display windows are designed to attract potential customers into the shop. In a busy shopping area, all the shop windows compete with each other to attract the attention of people passing by. They usually do this by using high light levels. However, a negative effect of this is a relatively high cost to the retailers, as the energy required to light the window display is high. Studies have been undertaken to investigate the use of coloured **LED**s for the background alongside bright light to focus attention on particular aspects of the display. This maintains the visual appeal of the shop window and reduces the lighting costs for the retailer.

You will see, in the film, a project to improve the lighting conditions in a retail outlet. The existing lighting system was dated and, as such, running costs were also quite high. The lighting designer noted that there were a lot of different light sources in the ceiling which made the area look untidy, and the merchandise seemed to disappear into the background due to the ineffective lighting.

The lighting designer decided to use different colour temperatures of lamp to make the area more inviting and make customers feel more at home, thus ensuring they stay longer in the shop. You will hear her explain how red looks warm and inviting, compared to the colder colours that are

Original lighting effect in bedding department

Final lighting effect in bedding department

often used in fast-food outlets (to encourage you to eat quickly and leave). **Colour temperature** is the colour appearance of the lamp itself and the light it produces. The colour temperatures of lamps make them visually 'warm', 'cold' or 'neutral' light sources.

**MATHS & SCIENCE**

**Colour temperature is measured on the Kelvin (K) scale. Kelvin is the basic unit of temperature.**

When specifying colour characteristics for a lamp, numerous psychological factors must be considered, depending on the lighting goals for the space. Warm light sources are generally preferred for the home, restaurants and retail applications to create a sense of warmth and comfort, while neutral and cold sources are generally preferred for offices and similar applications to create a sense of alertness.

In addition, in retail applications, colour is a critical design decision because buyers need to be able to choose products of the correct colour, both to enhance the chance of its sale and to reduce the chance of it being returned once the buyer gets outside and sees it under daylight. In this or any other application where the occupant needs to see the right colour, good colour quality is essential.

Lamps with a lower colour temperature have a warm or red-yellow/orangish-white appearance. The light is saturated in red and orange wavelengths, bringing out warmer object colours such as red and orange more richly. Lamps with a higher colour temperature have a cold or bluish-white appearance. The light is saturated in green and blue wavelengths, bringing out cooler object colours such as green and blue more richly.

If you refer back to the image of the retail outlet, you notice the almost 'cold' appearance and the way the colour of the fabric is lost. It is also difficult to distinguish the signage.

The final lighting design emphasises the colours of the merchandise. **Diffuse** lights have been used which distribute the light in an even manner.

**diffuse** – spread out and not directed in one place

Below are images from a different part of the retail store. Again see how the focus has been taken away from the ceiling and how the colours of the merchandise have been **accentuated**. This is known as **colour rendering** and describes how a light source makes the colour of an object appear to the human eye. The second image shows how subtle variations in the colour shades can be revealed.

# Maintenance

All lighting installations need to be maintained. This involves cleaning the lighting equipment, lamps and luminaires to ensure that light is not being wasted through obstruction by dirt. Also, since most lighting installations depend on a certain amount of reflected light from the room surfaces, these also need to be cleaned or redecorated periodically. The regularity of cleaning will depend on how dirty the environment is.

The question of maintenance is an important one. The lighting design engineer will need to assess what the reduction of light will be with respect to time due to dirt build-up and due to the reduction of lamp light output as the lamp ages. Correct assessment of these elements will enable the engineer to ensure that the required task illuminance is provided at the end of the **maintenance cycle**.

If this is not assessed correctly, there could be an excess provision of light and energy will be wasted. Equally, if the client does not carry out the recommended maintenance programme, they will be paying for light that is not provided, and the occupants will not have the required lighting conditions to ensure optimum performance.

To properly manage an existing system, many types of professionals may be involved, from electrical contractors to facilities managers. These personnel must ensure that the existing lighting system consistently provides the most cost-effective lighting at the lowest operating and maintenance cost. In the long term, this may entail upgrading the system to reduce energy costs and/or increase performance. However, on an ongoing basis, this involves a planned maintenance programme to keep the system operating at peak performance, and other activities that will ensure that the lighting system is continuously doing its job.

**MATHS & SCIENCE**

**The colour rendering index is a mathematical formula describing how well a light source's illumination of eight sample patches compares to the illumination provided by a reference source.**

**NUTS & BOLTS**

**To find out more about the lighting design of other spaces go to www.designlights.org/guides.html**

**accentuated** – made more noticeable

**maintenance cycle** – series of maintenance events that are regularly repeated in the same order

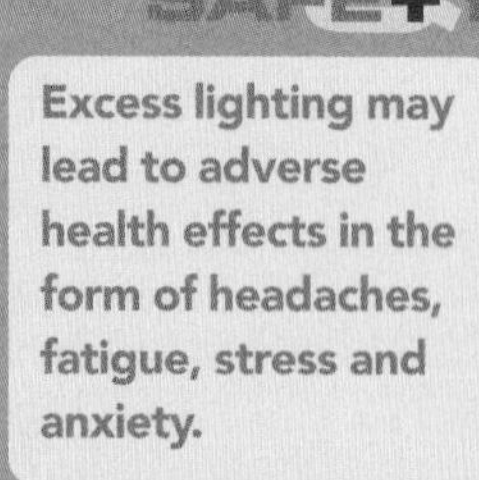

**HEALTH & SAFETY**

**Excess lighting may lead to adverse health effects in the form of headaches, fatigue, stress and anxiety.**

# The career path

## PERSONAL PROFILE: Faye Robinson

**Qualifications**

MSc Light and Lighting – focus on architectural lighting
BA (Hons) Theatre Design and Technology

**Professional experience**

*Lighting Designer* *Troup Bywaters + Anders*

- Principal lighting designer.
- Responsible for providing design and regulatory advice to all groups within the practice.
- Production of detailed design drawings and specifications, daylight modelling, and controls integration.
- Arranges in-house and external seminars.

**Projects**

The design and specification of a range of interior and exterior projects for different building types and client businesses, including retail, office environments and health care.

- Selfridges
- The Spinnaker Tower (Halcrow)
- Royal William Yard (Gillespies)
- Middlesbrough town centre regeneration
- Birmingham University
- Jasper Conran
- The Hacienda

### Journey to a lighting designer

With her background in scenography, Faye has been involved with the design and realisation of several theatrical productions, ranging from dance collaborations, musicals, improvised and outdoor theatre through to the more traditional productions. She continued to work within theatre whilst studying, providing lighting designs for UCL's Bloomsbury Theatre.

After completing her MSc, she began her career with Equation Lighting Design before joining the design department of the manufacturer Iguzzini as a specialist lighting designer.

**To find out more about a career in lighting design go to www.cibse.org and www.learndirect-advice.co.uk/helpwithyourcareer/ and www.tbanda.com**

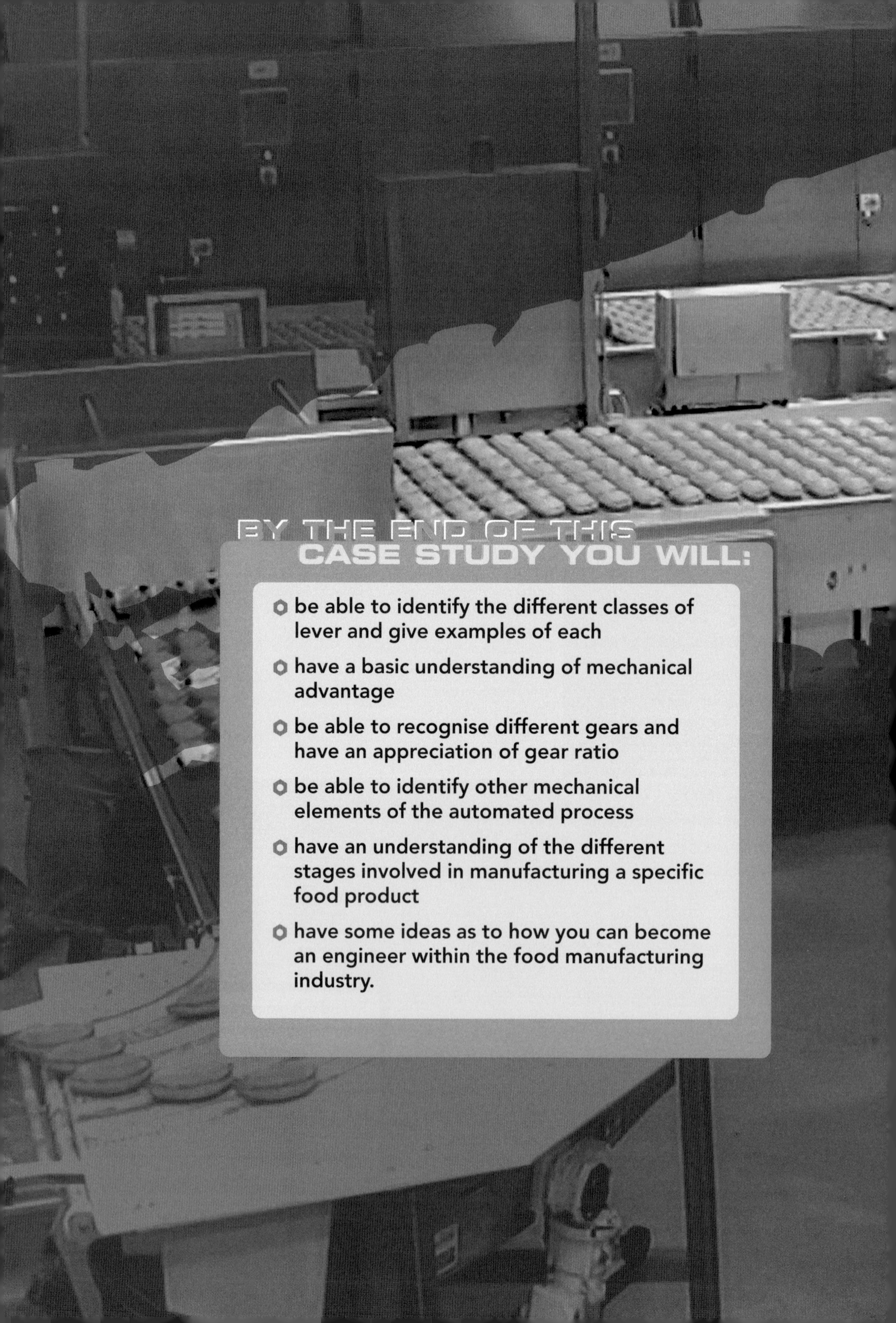

## BY THE END OF THIS CASE STUDY YOU WILL:

- be able to identify the different classes of lever and give examples of each
- have a basic understanding of mechanical advantage
- be able to recognise different gears and have an appreciation of gear ratio
- be able to identify other mechanical elements of the automated process
- have an understanding of the different stages involved in manufacturing a specific food product
- have some ideas as to how you can become an engineer within the food manufacturing industry.

# FOOD MANUFACTURING: GINSTERS

## Case study overview

This case study introduces you to the food manufacturing industry.

Over the following pages you will become familiar with a real company that manufactures food products by utilising a fully-automated manufacturing process.

You will examine the basic principles of levers and identify the elements needed for robotic movement. You will then follow the company's processes from the raw materials stage through to the final food product. This will help you to understand all the steps involved, the order in which they must be taken and the impact of automation.

Finally, you will be introduced to an engineer from the company who will describe the route he took to his current position.

# The sector

## Food and drink manufacturing

The food and drink manufacturing industry actually comprises over 30 different industries. These range from slaughterhouses, sugar refineries and grain mills to malt manufacturers and whisky distilling.

The UK is the world's fifth-largest grocery retail market and the largest in Europe, worth around **US$**156 billion in 2005.

The food and drink industry is the UK's largest manufacturing sector. With sales of around £74 billion, it accounts for 14 per cent of all manufacturing **gross value added**. Almost 7,000 food and drink companies employ 417,000 people – around 14 per cent of the manufacturing workforce.

**US$** – American dollars
**gross value added** – the increase in worth without any tax deduction
**collaboration** – working with someone to produce something
**strategic** – helping to achieve a plan

## Science-based innovation

The UK food industry has a reputation for innovation, and currently spends around £80 million a year on research and development. Its product development is aided by close **collaboration** with the large supermarket chains that dominate food distribution. These are forecast to have a 91 per cent share of the grocery market by 2010.

The government's Food Industry Sustainability Strategy, published in 2006, aims to encourage collaboration between industry and the UK science base. Government funding for food science and innovation has been increased by a third to support **strategic** research and knowledge transfer, focused on manufacturing efficiency and high-quality, healthy food products.

NUTS & BOLTS

**To find out more about the food manufacturing industry go to www.improveltd.co.uk**

## Career choice

The government wants the food industry to be promoted as an employer of choice. As such, the standard of workforce training has been enhanced through the national network of Centres of Vocational Excellence. Over 40 of these now cover food technology, manufacturing, hospitality and catering.

# The company

## Background

The Cornish pasty has existed for over 300 years, and was initially eaten by fishermen, miners and farm workers as a snack to eat away from home. The pastry case enabled the meat and vegetables to remain clean and warm throughout the day, and also helped to retain the rich flavour of the filling.

A traditional Cornish pasty, like the Ginsters pasty, has a crimp on the side rather than along the top, which is a characteristic of the Devon pasty. Around 1945 through to1950, the popularity of the Cornish pasty began to spread as holidaymakers ate the product as a snack on the beach or while motoring. Demand for the product from outside the Cornwall area began to grow, and many manufacturers started to produce 'pasties' around the country, under the guise of being genuine Cornish pasties. In certain parts of the country, poor-quality product gave rise to a bad reputation for the humble pasty.

In 1969, Geoffrey Ginster began producing his own Cornish pasties from a site in Callington in Cornwall. This was just two years after starting a van-sales business buying and selling fresh pasties to local retailers in Cornwall. The success of the business quickly began to grow as **distribution channels** were opened up throughout the south of England. Eventually, distribution progressed so rapidly that the quality Cornish pasty could be enjoyed throughout the UK.

In 1977, Samworth Brothers acquired Ginsters. With the support of Samworth Brothers, both the brand and the **product portfolio** have grown from strength to strength by targeting the ferocious appetite of the UK snacking market.

**distribution channels** – the manner in which goods move from the manufacturer to the points of sale

**product portfolio** – the set of different products that a company produces

## Modern day

Ginsters Original Cornish Pasty has now grown to become the nation's biggest-selling product in the chilled savouries market. The brand as a whole now has a turnover of £140 million at retail selling prices. As well as traditional favourites such as pasties, pies and sausage rolls, the brand range includes deep-fill sandwiches, savoury wraps and innovative snacks, such as buffet bar, roaster and pasta salads. The brand is now the biggest advertiser in its category, and Ginsters is supported by a national poster, magazine and radio campaign on the theme of 'Cornish Through and Through'.

Today, the company employs 700 people and has its own fleet of 150 vehicles based at sales offices across the country. They deliver direct to motorway service stations, forecourts, convenience stores and foodservice customers. The Ginsters range is also available from the major high-street **multiples**.

In terms of current production, Ginsters produces 3.4 million units (pasties) per week. The future target has been set at four million per week.

# The product

## Introduction

**multiple** – a shop with branches in many places
**outlay** – an amount of money spent on something

Whilst the actual product is a pasty, we are going to consider the engineering processes and equipment necessary to enable production to take place.

Ginsters uses a wide range of robots, as seen in the film, at all stages of the manufacturing process. Most of the robots used in engineering are used to lift heavy objects or for repetitive manufacturing. They can also be used where tasks are difficult or dangerous.

Typical applications for industrial robots include welding, painting, assembly, pick and place, packaging, product inspection, and testing.

While a robot works relentlessly and without breaks, they are very costly to install. Ginsters will take over four years to recover the large **outlay** needed to install some of their robots.

Robots are, however, excellent for food manufacturing as they perform repetitive tasks efficiently with speed and precision, and reduce contamination of the food which can be caused by bacteria from human hands.

## Robotics

A robot needs an input, process and output to work. They also need feedback.

- The input takes the form of a control instruction.
- The process is the action performed by the robot.
- The output is the completed task.

FIG 6.1: Robotic system

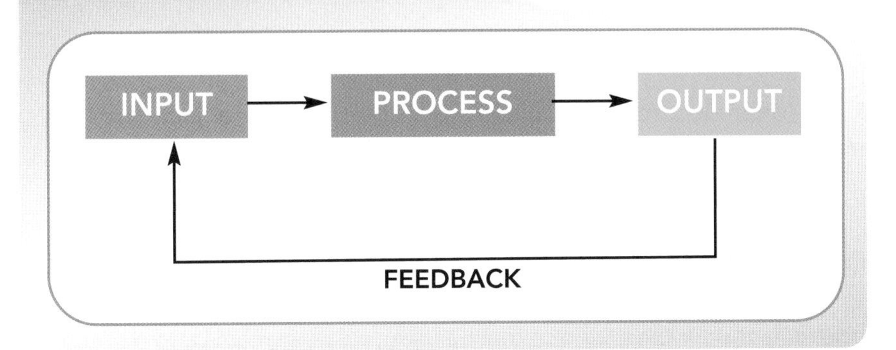


- Feedback is the information from the sensors that helps the controlling computer to issue the correct input instructions.

All robots need sensors to measure their movement, and some type of **actuator** to move them so that they can carry out a process. Two main movement systems are used to move a robot: stepper motors and hydraulics.

**actuator** – a device that creates mechanical movement

The region of space that a robot can reach is called the **working envelope**.

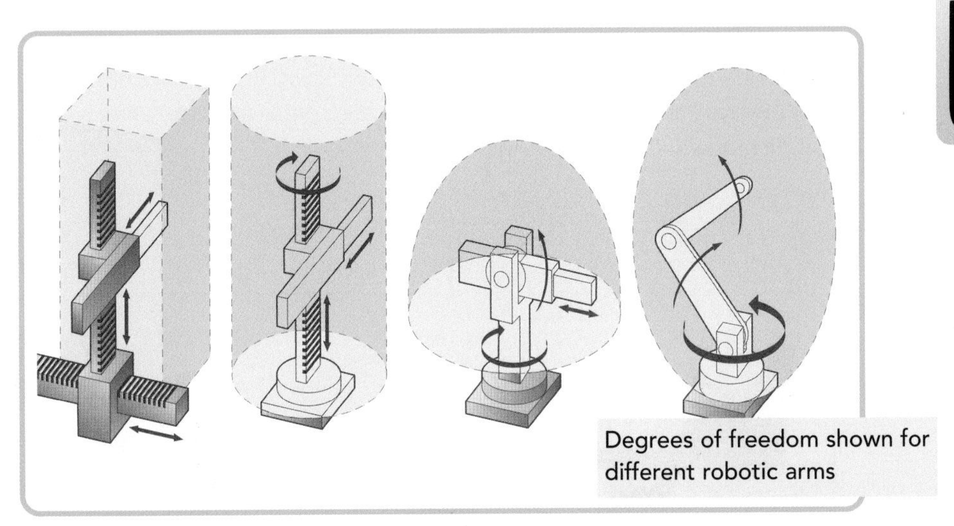

Degrees of freedom shown for different robotic arms

The number of degrees of freedom determines the working envelope.

Most robots use a robotic arm. Robotic arms usually have six degrees of freedom, which means that they can pivot in six different ways.

The purpose of all the robotic arms shown in the film is to move an object from one place to another: for example, picking up the pasty and packing it in a box. At the end of the robot arm there is an actuator, which works in a similar way to that of the human hand in that it can grab and carry different objects. The hand can contain all sorts of built-in sensors to provide feedback, including pressure sensors that can tell the computer controlling the robotic arm how hard the robot is gripping the object.

Robotic arm used by Ginsters

# Force

To understand how the robot works an engineer first needs to understand forces.

Force (often denoted **F**) is a push or pull exerted on an object. The **magnitude** of the force is a measure of how hard the push or pull is. In addition to magnitude, force has a direction.

Forces are applied to the robotic arms by **leverage**.

A lever is the simplest kind of mechanism, which enables us to do work more easily. Levers are used to lift heavy weights with the smallest amount of effort.

All levers have a **fulcrum**, an **effort** and a **load**.

There are three classes of lever.

## Class 1 lever

A class 1 lever has the load and the effort on opposite sides of the fulcrum, like a see-saw.

If two identical twins sit at either end of a see-saw, they will remain **static**. This is because the see-saw is balanced in the centre and the twins are an equal distance away from the fulcrum.

**magnitude** – size or extent
**leverage** – exertion of force by means of a lever
**static** – staying in one place without moving

**MATHS & SCIENCE**

We use levers in everyday life. Bicycle brakes work due to the fact that they are based on levers.

**MATHS & SCIENCE**

The fulcrum (or pivot) is the point where the load is pivoted. The effort is the force applied to move the load.

**MATHS & SCIENCE**

When two turning forces are equal and opposite, they are balanced, there is no change in the movement of the system, and the system is said to be in equilibrium.

**FIG 6.2:** See-saw in equilibrium

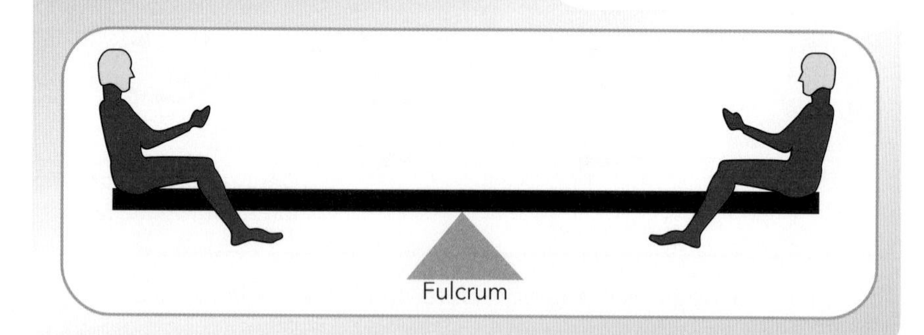


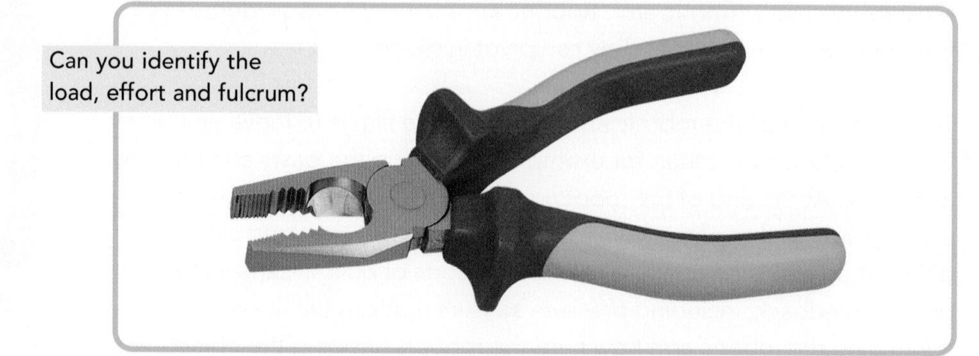

Can you identify the load, effort and fulcrum?

## Class 2 lever

A class 2 lever has the load and the effort on the same side of the fulcrum, with the load nearer the fulcrum.

FIG 6.3: Class 2 lever

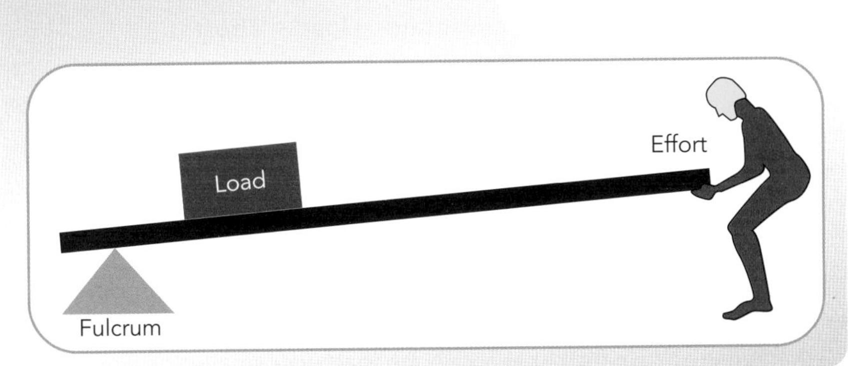


An example of a class 2 lever is a wheelbarrow.

Can you identify the load, effort and fulcrum?

## Class 3 lever

A class 3 lever has the load and the effort on the same side of the fulcrum, but the effort is closer to the fulcrum than the load, so more force is put into the effort than is applied to the load.

FIG 6.4: Class 3 lever

An example of a class 3 lever is a pair of tweezers. Another example is the human arm or robotic arm.

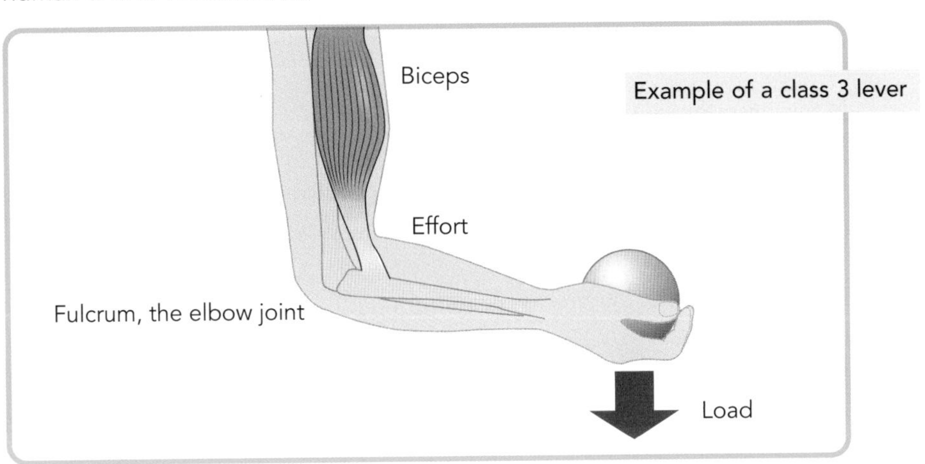


Example of a class 3 lever

# Mechanical advantage

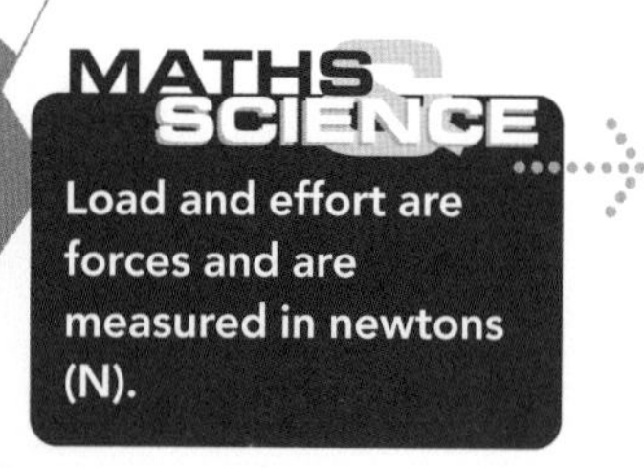


Load and effort are forces and are measured in newtons (N).

Class 1 and class 2 levers both provide mechanical advantage, which means that they allow you to move a large load with small effort.

**Mechanical advantage = load ÷ effort**

A class 3 lever has mechanical disadvantage as the effort is always greater than the load – the force needed to use them is greater than the force that they can move. However, the distance moved by the load is greater than the distance moved by the effort.

With class 1 and 2 levers you have to move the effort a greater distance to enable you to move the load a short distance. That is, you need to push the lever down further to move the load up a smaller distance.

FIG 6.5: Class 3 lever demonstrating distances moved through

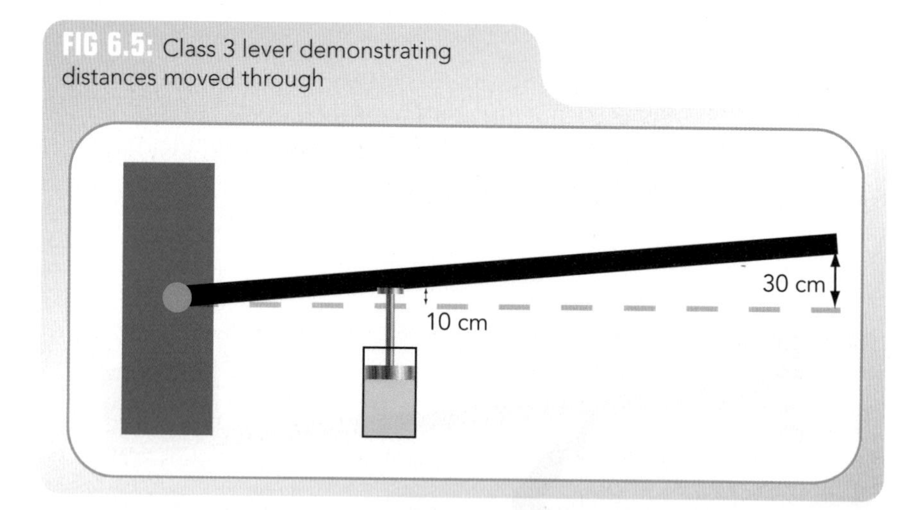


Look at Figure 6.5. If the effort (hydraulic cylinder) moves up 10 cm, the end of the lever will move 30 cm over the same time period. The distance and velocity ratios are therefore the same and can be calculated by the formula below.

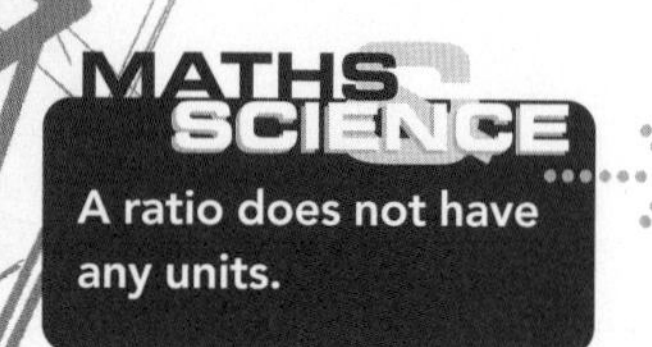


A ratio does not have any units.

$$\text{velocity ratio} = \frac{\text{distance moved by effort}}{\text{distance moved by load}}$$

In this case the velocity ratio is 1/3.

The closer the effort is to the fulcrum the greater the distance the load moves through, but the load the system can move is smaller.

The fact that class 3 levers apply more force to the effort rather than the load means that these levers are good for grabbing something small, fiddly or dirty, or picking up something that could be squashed or broken if too much pressure is applied. These are the exact properties needed by the robotic arms used by Ginsters.

# Moving the robotic arm

Robotic arms can be moved by stepper motors or by hydraulics.

## Stepper motor

A stepper motor is a digital device. Digital information is processed to achieve the type of controlled motion needed by robotic arms. A stepper motor follows digital instructions provided by the computer.

Basically, stepper motors are electrical motors that are driven by digital **pulses** rather than a continuously applied voltage. Inherent in this concept is **open-loop control**, wherein a train of pulses **translates** into so many shaft **revolutions**, with each revolution requiring a given number of pulses.

Each pulse equals one **rotary** increment, or step, which is a portion of one complete rotation. By counting the pulses applied, it is possible to achieve the desired amount of rotation.

The precision of the controlled motion is determined by the number of steps per revolution; the more steps, the greater the accuracy that can be achieved.

## Speed of movement

To achieve the required speed, Ginsters had to redesign the robotic arms using carbon fibre to make them lighter (reduce mass) and therefore easier to move.

## Gears

Robotic arms use gears alongside levers to achieve mechanical advantage. Gears consist of toothed wheels fixed to shafts. The teeth interlock with each other, and as the first shaft rotates, the motion is transmitted to the second. The motion output at the second will be different from the motion output of the first.

Gears are generally used for one of four different reasons:

- to reverse the direction of rotation
- to increase or decrease the speed of rotation
- to move rotational motion to a different **axis**
- to keep the rotation of two axes **synchronised**.

## Gearing ratios

Where there are two gears of different size, the larger gear will rotate more slowly than the smaller gear. The difference between the two speeds is called the velocity ratio, or the **gear ratio**.

**pulse** – a brief sudden change in a normally constant quantity
**open-loop control** – a system controlled directly and only by an input signal without feedback
**translate** – convert something or be converted into
**revolution** – the single completion of an orbit or rotation
**rotary** – rotational; revolving around a centre
**axis** – a real or imaginary straight line which goes through the centre of a spinning object
**synchronised** – happening at the same time

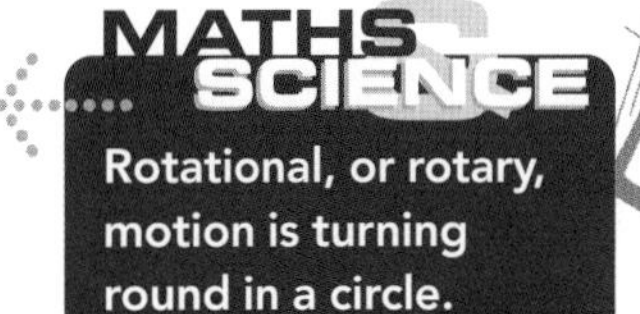


Rotational, or rotary, motion is turning round in a circle.

**MATHS & SCIENCE**

A diameter of a circle is a straight line passing through the centre which touches the curve forming the circle at each of its ends.

For any gear, the ratio is determined by the distances from the centre of the gear to the point of contact. For instance, in a device with two gears, if one gear is twice the diameter of the other, the ratio would be 2:1.

The number of teeth on a gear wheel can be used to determine exact gear ratios. You just count the number of teeth in the two gears and divide. So if one gear has 60 teeth and another has 20, the gear ratio when these two gears are connected together is 3:1.

$$\text{Gear ratio} = \frac{\text{number of teeth on driven gear}}{\text{number of teeth on driving gear}}$$

The higher the gear ratio the more leverage a machine has.

## Gear trains

To create large gear ratios, gears are often connected together in **gear trains**, as shown here.

FIG 6.6: Gear train

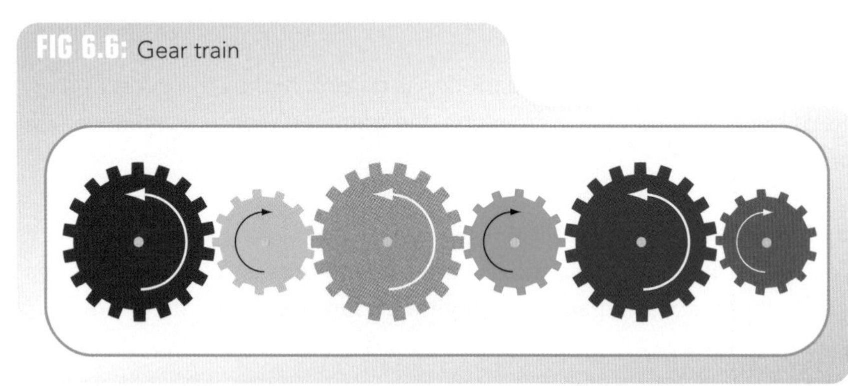

## Compound gears

A compound gear is where you have two gears fixed on the same shaft.

FIG 6.7: Compound gear

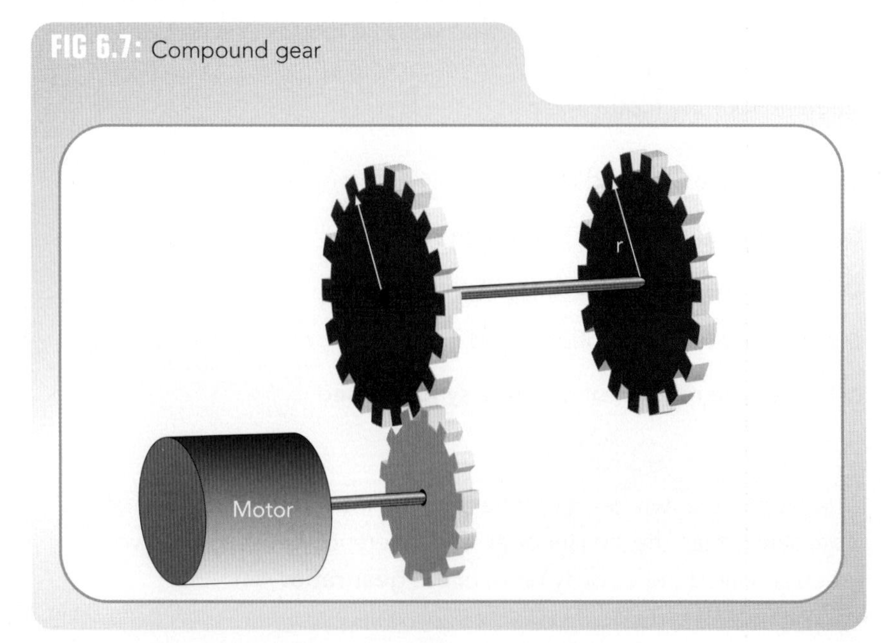

When two gears are on the same shaft, the speed of rotation is exactly the same.

A **compound gear train** is one which has two or more gears attached to the same shaft. In actual fact, it is a combination of two or more gear trains. Compound gear trains are used where large speed reductions are required.

## Speed

If you know the gear ratio and the speed of input, you can calculate the speed output using the formula:

**Output speed = input speed ÷ gear ratio**

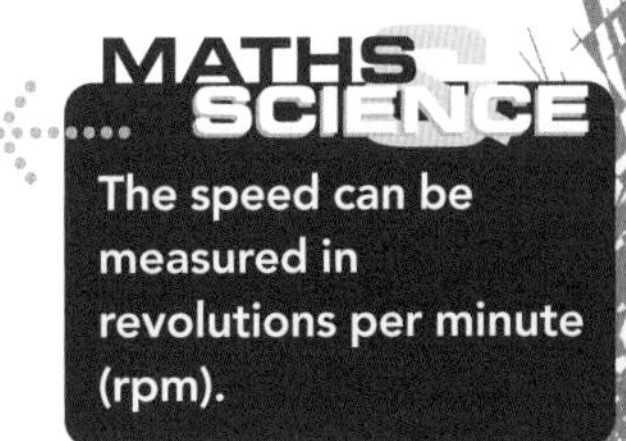


## Gear pitch

When selecting gears, an engineer must consider **pitch**, **pitch diameter**, and number of teeth.

The pitch of a gear is the distance between equivalent points on adjacent teeth. For two gears to mesh they must have the same pitch.

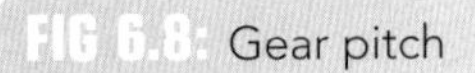
FIG 6.8: Gear pitch

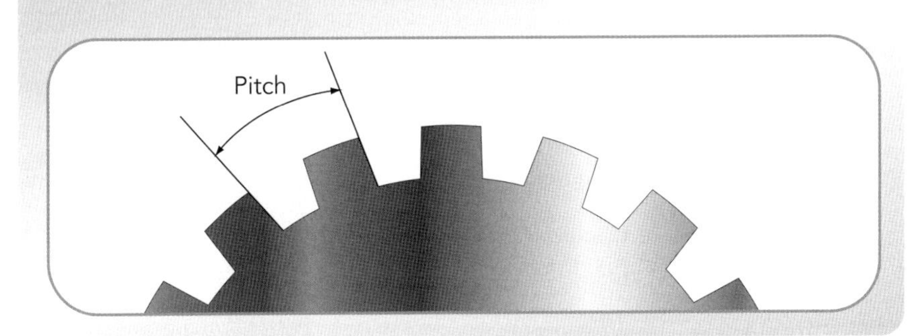


The pitch circle of a gear is an imaginary circle which passes through the point where the teeth touch when one gear meshes with another. The pitch diameter is the diameter of the pitch circle.

FIG 6.9: Pitch circle of a gear

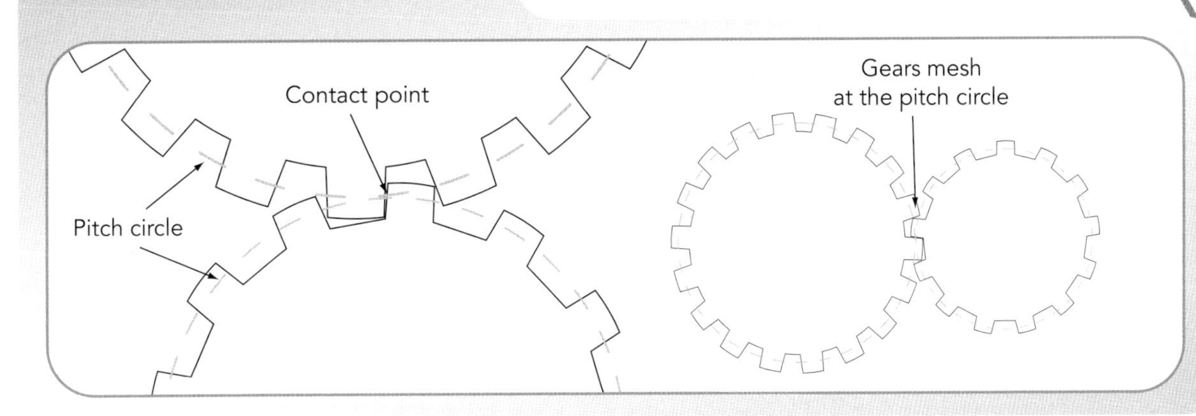

**MATHS & SCIENCE**

The symbol ≈ means 'is approximately equal to'.

**helix** – something spiral in form

**MATHS & SCIENCE**

Linear motion is moving in a straight line. Reciprocating motion is moving backwards and forwards in a straight line.

## Types of gears

- **Spur gears** (≈90% efficiency)
  - Most commonly used gears.
  - Inexpensive to manufacture.
  - Not recommended for very high loads as gear teeth can break more easily.

- **Helical gears** (≈80% efficiency)
  - Operate just like spur gears, but offer smoother operation (due to continuous tooth mating).
  - Have a higher load capacity (as the teeth have a greater cross section).
  - More expensive to manufacture.

- **Bevel gears** (≈70% efficiency)
  - Straight bevel – these are like spur gears; the teeth have no **helix** angle.
  - Spiral bevel gears – teeth have a spiral angle, which gives performance improvements much like helical gears.

- **Rack and pinion** (≈90% efficiency)
  - Change linear motion into rotary motion and vice versa.

- **Worm gears** (≈70% efficiency)
  - The low number of teeth can result in a very large velocity ratio.
  - Can carry high loads.
  - Because of sliding action, efficiency is low.

# Moving the product to the robots

In the Ginsters film, you will notice that the pasties are moved from each operation by a range of conveyors.

The simplest type of conveyor uses the weight of the object to move it. They consist of a set of rollers. You may have used a conveyor like this at the airport where small bags are fed into and out of the X-ray machines at airport security.

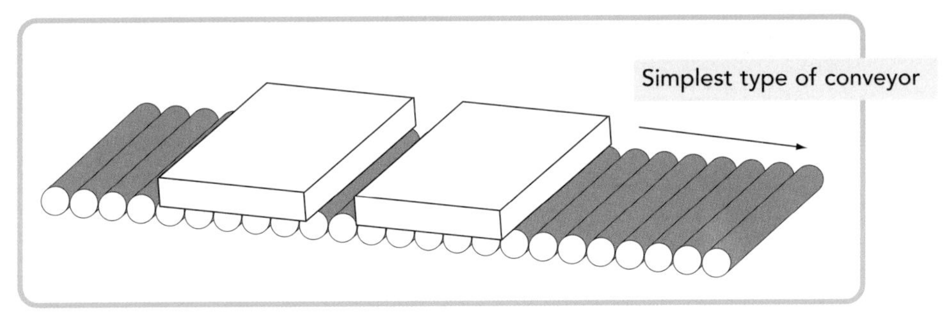


Most conveyors used in engineering are powered. They consist of a bed of set length and width, with a shaft and roller at each end.

**FIG 6.10:** Basic elements of a powered conveyor

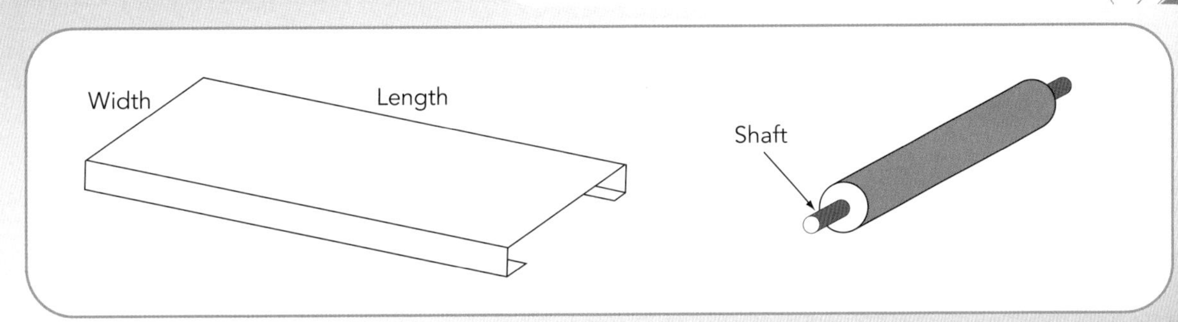


Next a motor is added, with a drive chain or gears.

**FIG 6.11:** A motor and drive pulley

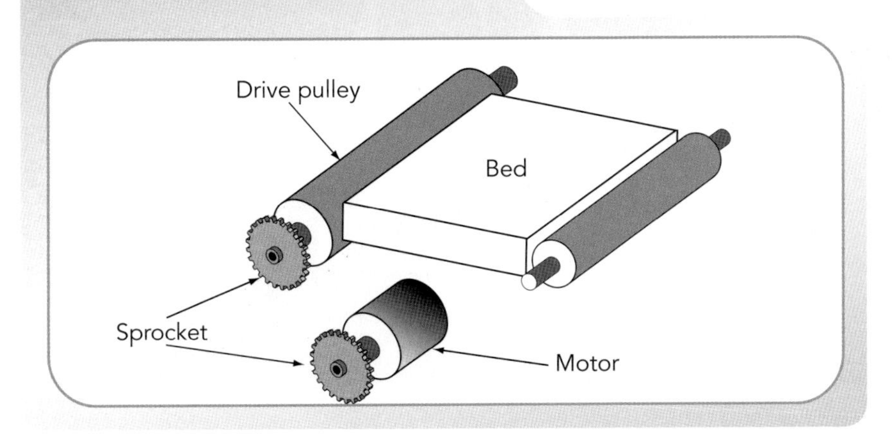


**FIG 6.12:** Conveyor belt held tight by take-up screw

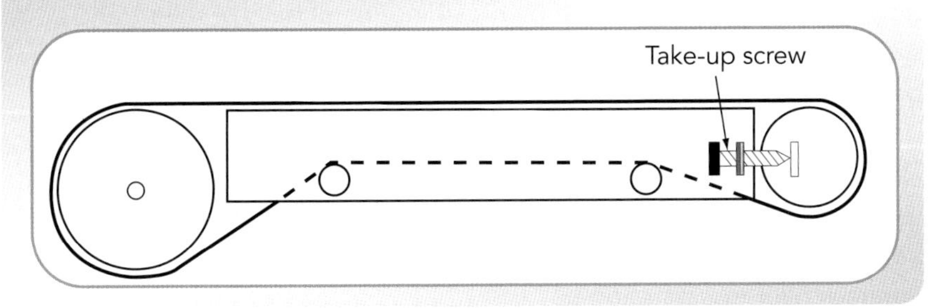

**tensioning** – tightening

A belt is added with some sort of **tensioning** screw to keep the belt tight. There are a wide range of materials used for belts. Where they are likely to stretch a large amount, the take-up screws are replaced with a take-up roller mechanism or adjustable rollers called snub idlers.

Some conveyors do not use a belt – the rollers are themselves powered.

**HEALTH & SAFETY**

**Many of the conveyors used in food manufacturing have special antimicrobial belts to inhibit the growth of microbes such as bacteria.**

FIG 6.13: Conveyor with powered rollers

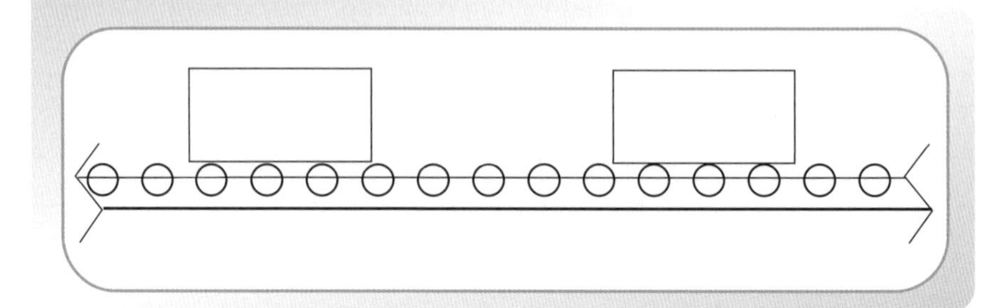

Powered conveyors are essential for long horizontal transport, for applications with significant inclines or declines, and where a high degree of automation is required, such as merging, diverting and sorting.

The conveyors you see in the film are used for the transportation of the pasties from one level to another, and for horizontal transport.

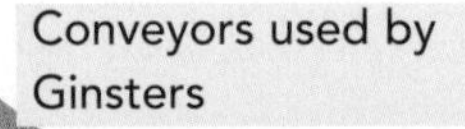

Conveyors used by Ginsters

## Bearings

The concept behind bearings is very simple. Things roll better than they slide. Bearings reduce friction by providing metal balls or rollers and a smooth inner and outer metal surface for the balls to roll against. These balls or rollers bear the load, allowing the device to spin smoothly.

Ball bearing

There are many types of bearings used in the Ginsters machinery, each used for a different purpose. The main types of bearings are ball bearings, roller bearings, ball thrust bearings, roller thrust bearings and tapered roller thrust bearings.

**Ball bearings** use metal balls housed within a case called an outer race. The other moving part is connected to an inner race. Because the balls have a small point of contact they spin very smoothly. However, as there is little contact it also means that there is not very much contact area holding the load, so if the bearing is overloaded, the balls deform, ruining the bearing.

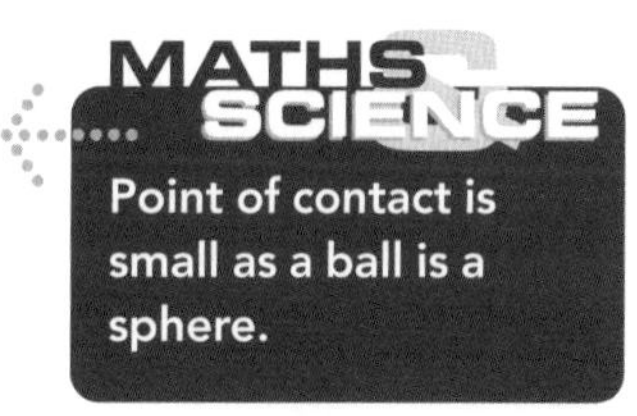


Ball bearings are usually the least expensive choice, provided loads are light, giving them a good price/efficiency ratio.

**Roller bearings** are used in conveyor belt rollers, where they must hold heavy loads. The roller is a cylinder, which means the load is spread out over a larger area, allowing it to handle much greater loads than a ball bearing. Needle bearings are like roller bearings but have a very small diameter roller. They are used when there is only a small space to fit the bearing into.

Roller bearing

# The process

## Planning

Ginsters uses a wide range of automation and robotic equipment. However, robots are not as flexible as humans, and products handled by them must be a set size. In many production lines this is simple to achieve, but the fact that the pies and pasties have to be cooked causes a problem.

Cooking causes shrinkage. The amount of shrink will depend on the mix of the pastry and filling. Any slight change in consistency will affect the overall size. To ensure consistency in the pastry, you will hear the engineer in the film refer to measuring not just water, fat and flour but the amount of energy put into the product by the mixer.

MATHS & SCIENCE

Energy is the capacity to do work. When work is done on or by an object, it gains or loses energy respectively.

Designing and running an automated production line when you are working with cooked products is not easy, particularly when you are using pastry products such as puff pastry, which can expand irregularly if it is not made to an exact consistency.

## Production

### Automating the cooking and packing process

There are two types of ovens that can be used in commercial food production: in-line ovens and rack ovens. In the film you will see the use of in-line ovens, which have a continuous steel-band conveyer which carries the product through the oven.

Ovens used in the Ginsters factory

**HEALTH & SAFETY**

**In the food and drink industries, about 600 injuries per year are reportable to the Health and Safety Executive.**

Different areas of the oven are heated to different temperatures. The pasties are heated to a core temperature of over 98 degrees Celsius. The speed of the steel belt is crucial: too fast a speed would result in the product being undercooked, and too slow a speed would mean that the product would be overcooked. The average cooking time of the pasty is 24 minutes, and approximately 7,800 units per hour enter the ovens.

The conveyors in the ovens have to function at high temperatures. The bearings and steel belts have to be specially designed to function efficiently at such high temperatures.

It is worth noting here that in the event of a fire alarm sounding, the oven must continue to work, as the band cannot be stopped without destroying all the products within the oven.

When selecting an in-line oven, consideration needs to be given to the number of products that can fit across the band as this determines the speed of the whole production line.

Rack ovens are used where there are short production runs. They work in the same way as a conventional oven, where the product is placed in the oven on large racks. Sometimes rack ovens can be faster than in-line ovens, as they can often hold larger quantities of the product at any one time.

Once the product has been cooked, it needs to be cooled as quickly as possible to reduce the time that bacteria has to grow, and so that it can be packaged. The cooling machines, called spirals, are circular conveyor belts. There are two cooling spirals shown in the film. The first reduces the pasty's core temperature from 98 to 45 degrees Celsius in 45 minutes. The second spiral, with a temperature of -24 degrees Celsius, takes

35–40 minutes to reduce the core temperature of the pasty to less than 5 degrees Celsius on the outfeed.

The spirals themselves are a magnificent piece of engineering. Conveyor spirals and bearings have to function at very low temperatures. Original spirals had to be defrosted regularly, and this meant that the production line had to stop.

The new spiral machines contain an **autochanger**, allowing one part to be defrosted whilst the other part continues running.

**autochanger** – mechanism for the automatic substitution of one thing for another

Spirals cool the pies by blasting cold air, at very low temperatures, over the product. After the pasty is cold, it can be packaged in the plastic wraps, then in boxes. In the film you will see a 'place robot'.

First the pasty passes through what is described in the film as a vision system. The vision system has sensors that determine each pasty's length, width and height, colour and decoration. If it passes these automatic vision tests, the robots place the pasty on an infeed conveyor leading to the flow wraps. Here the product is heat-sealed into film.

HEALTH & SAFETY

**Conveyors are involved in 30 per cent of all machinery accidents in the food and drink industries – more than any other class of machine.**

Of course, it is not just the robots that require an accurate-size product to enable them to perform efficiently: the preformed flow wrap packaging is also designed to fit the product fairly tightly. Packaging comes preprinted and to a set size. It is critical that the product (in this case a pasty) fits into the packaging so that the picture and graphics are in the correct place. If the margins are not correct, the product itself looks less appealing.

The product is packaged in a continuous-feed wrapping process. The wrapping is cut after the product has been fed in. If the product is too large, there is a risk of 'chops' – this is where the product itself is cut.

Another problem which can occur is where the wrapping is not sliced correctly, leading to a large 'snake' where the film wrap has not separated and a large number of products are in one long wrapping. When this long snake reaches a turn in the conveyor, it can cause a blockage, potentially wasting a large number of products.

Once wrapped, the pasties are then carried by another conveyor to a secondary packing area where they are packed into boxes. The boxes travel down yet another conveyor to the palletising system, where three robots pick and collate the product.

The robot picks up the boxes and places them on a pallet. An automatic wrapping machine clingfilm-wraps the pallet, which then travels down a conveyer system to the warehouse ready for dispatch.

## Packing areas

There are a number of packing areas shown in the film, and these can be split into two main safety areas: low-care and high-care.

Low-care and high-care packing areas

By the time the product reaches the low-care area it is already packed. Consequently the hygiene rules are less strict.

In the high-care area, the product is in its naked state and can be touched by human hands. At this stage there are very strict hygiene rules, as any contamination would be critical.

To make sure that the right people are in the right areas, different coloured coats are worn.

# Testing and minimising waste

One of the biggest issues for Ginsters is minimising the amount of waste produced. There will always be some waste generated, as products need to be tested. This is done using temperature probes and by breaking apart the product to ensure the inside is of the correct cooked consistency. Any products tested in this way need to be destroyed.

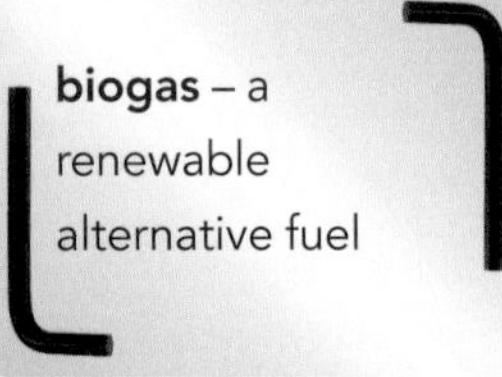

**biogas** – a renewable alternative fuel

Products that are slightly too large or too small can, in the case of Ginsters, be sold to staff, but products which are broken or damaged have to be scrapped. Other scrap includes potato peelings and the fat removed from meat. To reduce the environmental impact of this waste, Ginsters send scrap products to a **biogas** plant to be turned into methane gas. This is a form of waste exchange – the waste product of one process becomes the raw material for a second process.

**MATHS & SCIENCE**

**Methane is a colourless, odourless inflammable gas. Natural gas contains 99% methane.**

Ideally there would be no waste generated, as this has a cost impact on the company. The later the waste is generated on the production line, the more costly it becomes. A company is reluctant to invest energy and time into adding value into any raw materials only to have to scrap the output.

# The career path

## PERSONAL PROFILE: Mark Alexander

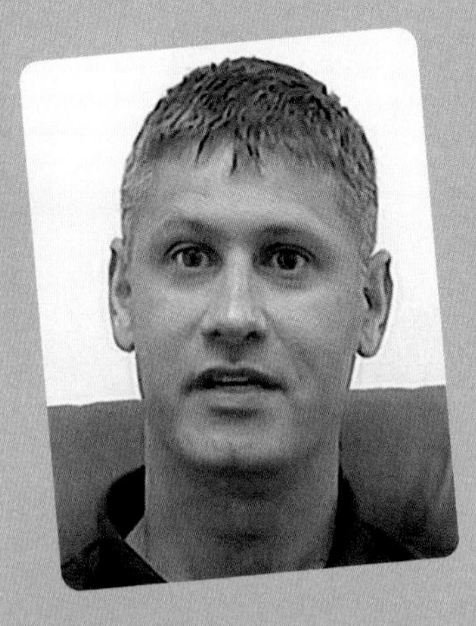

**Professional experience**

*Head of Engineering* — *Ginsters*

- Responsible for site engineering and projects.
- Responsible for ensuring site equipment, site services and production line equipment are maintained correctly.

*Engineering Manager* — *MD Foods (dairy company)*

- Reorganisation of the engineering department from single-skill base to multiskilled engineers.
- Responsible for a five-year investment programme of £13.5 million to increase site volume from 2.2 million litres a week to 8.4 million.

*Engineering Manager* — *Unigate Dairies*

- Joined as a Shift Technician.
- Involvement with the installation and commissioning of all site packaging and process equipment.
- Promotion to Team Leader after 2 years, and Assistant Engineering Manager after 3 years.
- Engineering Manager for 5 years and then headhunted to go to work for MD Foods.

### Journey to Head of Engineering

I left school aged 16 and joined Gravfil machines to complete a four-year indentured apprenticeship in the machine shop, manufacturing parts for filling and capping machines. After completing my City & Guilds I went on to complete an ONC in Mechanical and Production Engineering.

I left Gravfil machines at the end of my apprenticeship because I wanted to continue my studies. I joined Normand Electrical (a manufacturer of electric motors and gearboxes), who allowed me to continue with day release to complete my HNC. I worked for Normand Electrical for two years as a Planning Engineer. On completion of my HNC I left Normand Electrical and joined Unigate Dairies.

When I joined Unigate Dairies it was the most modern automated dairy in Europe and still under construction. This was the start of multiskilled technicians (electrical, mechanical, instrumentation) at Unigate.

**To find out more about a career in the food manufacturing industry go to www.prospects.ac.uk**

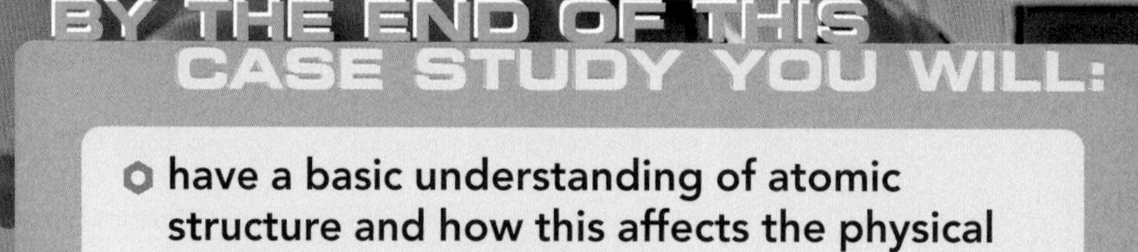

## BY THE END OF THIS CASE STUDY YOU WILL:

- have a basic understanding of atomic structure and how this affects the physical properties of a material
- be able to identify the main physical properties of ceramics
- have an appreciation of the planning and production stages involved in the manufacture of a plug and socket for a vacuum chamber
- have an understanding of the different tests undertaken to test the integrity of a seal
- have some ideas as to how you can become part of the precision engineering sector.

# PRECISION ENGINEERING: CERAMIC SEALS

## Case study overview

This case study introduces you to the world of precision engineering.

Over the following pages you will become familiar with a real company that specialises in the production of ceramic-to-metal seals.

You will examine the properties of ceramics through consideration of their atomic structures. You will then follow the company's processes from initial design to the final manufactured product, with particular emphasis on the testing processes.

Finally, you will be introduced to an apprentice from the company who will describe the route he took to his current position.

# The sector

## What is precision engineering?

Precision engineering involves work at the forefront of current technology. Present advanced technology products are dependent on high-precision manufacturing processes, machines, control technologies and even **nanotechnology**.

The achievement of ultra-high precision in the manufacture of extremely small devices widens the opportunities for technological advancement in several diverse and futuristic fields, such as massive computing power, global personal communication devices and high-resolution optical devices.

Precision engineering is cross-disciplinary, dealing with materials, machining processes, machine tool design, microsensors and actuators, manufacture of integrated circuits and mass storage devices, novel manufacturing methods, and applications in medicine and many other relevant fields.

The development of new high-quality products depends on the challenge of high precision.

**nanotechnology** – the branch of technology that deals with dimensions and tolerances of less than 100 nanometres
**turnover** – the amount of money taken in a particular period
**fragmented** – consisting of several separate parts
**metallurgical** – concerned with the properties of metals, their production and purification

## Engineering sector

The **turnover** of the UK engineering sector is around £90 billion. Exports account for around 40 per cent of this turnover (£35 billion). Machinery and equipment make up 40 per cent of total UK engineering exports.

The sector is exceptionally **fragmented**, comprising over 50,000 firms with an average turnover of £1.5 million, and an average workforce of just 18. It encompasses a broad spectrum of subsectors, including:

- Mechanical, electrical and process engineering – covering the complete spectrum of electrical, electronic and mechanical industries, from capital power equipment to wiring accessories.
- Metals and minerals – including steel, non-ferrous, ferrous and precious metals, glass, forging and foundries, metal stockholding, finishing and recycling, industrial ceramics and minerals, and sheet-metal working.
- **Metallurgical** process plant – including aluminium smelters,

non-ferrous metals plant and processing lines, pickling lines, steel mills and hot, cold and long product rolling.

- Mining – all establishments primarily engaged in mining, including quarrying, well operations, milling, and exploration and development of mineral properties.

NUTS & BOLTS

**To find out more about materials and applications of advanced technology go to www.materials-careers.org.uk**

# The company

## Background

Ceramic Seals Limited is an independent company that manufactures a comprehensive range of quality ceramic-to-metal seals, sapphire-to-metal seals and related products.

Formerly a division of Ferranti International plc, world leaders in this specialised technology, Ceramic Seals Limited has over 40 years' experience in the design and manufacture of ceramic-to-metal seals. These seals are used in high-vacuum, high-pressure, high-temperature and **cryogenic** conditions.

**cryogenic** – the branch of physics dealing with the production and effects of very low temperatures

## Modern-day applications

Ceramic Seals' product range includes electrical/fluid feedthroughs, isolators, cable terminations, and sapphire lenses and windows. All products are specifically designed and produced to meet the demands of, amongst others, the vacuum, nuclear, electrical, aerospace and defence industries.

**traceablility** – ability to find the origin of something

**procurement** – the obtaining of supplies

NUTS & BOLTS

**To find out more about Ceramic Seals go to www.ceramicseals.co.uk**

## Standards

The company is certified to BS EN ISO 9001: 2000 and is committed to maintaining a high level of quality in every aspect of its activities.

Certificates of conformity are often supplied to customers, and full **traceability** of the product is possible from raw material to finished product.

The company's commitment to quality is evident at all stages from **procurement** through manufacturing, testing and customer service, with careful monitoring to comply with documented specifications.

# The product

## Introduction

Engineers work with a wide range of engineering materials. The chart below shows some of these materials.

| Engineering materials | | | | | | |
|---|---|---|---|---|---|---|
| Metals | | Plastics | | | Ceramics | Others |
| Ferrous | Non- ferrous | Thermoplastics | Thermosets | Elastomers | | |
| Steels | Aluminium | Acrylics | Epoxies | Rubbers | Oxides | Reinforced plastics |
| Stainless steel | Copper | ABS | Phenolics | Silicones | Nitrates | Laminates |
| Tool and die steels | Titanium | Nylons | Polyimides | Polyurethanes | Carbides | Metal matrix |
| Cast iron | Tungsten | PVC | | | Glass | Ceramic matrix |
| | | | | | Graphite | |
| | | | | | Diamond | |
| | | | | | Glass ceramics | |

In this section of the book we will explore engineering ceramics. There are two types of ceramics: **traditional ceramics** and **industrial ceramics**.

Traditional ceramics are used to make things such as tiles, bricks, sewer pipes, pottery and abrasive wheels.

Industrial ceramics, sometimes called engineering high-tech ceramics or fine ceramics, are used to make things like turbines and automotive and aerospace components.

Ceramic Seals only works with industrial ceramics.

In general, ceramics have the following properties:

- hard
- resistant to plastic deformation
- resistant to high temperatures
- good corrosion resistance
- low thermal conductivity
- low electrical conductivity.

However, some ceramics exhibit high thermal conductivty and/or high electrical conductivity. The combination of these properties means that ceramics can provide:

- high wear resistance with low density
- wear resistance in corrosive environments
- corrosion resistance at high temperatures.

The choice of ceramic is based on the final application of the component to be manufactured, as well as the manufacturing process to be undertaken.

**ion** – an electrically charged particle
**electron** – particle with a negative charge and very small mass

## Atomic structure

The atomic structure of ceramic **crystals** is one of the most complex of all material structures. Ceramics are compounds of metallic and non-metallic elements.

Two types of bonding mechanisms occur in ceramic material, **ionic** and **covalent**. An ionic bond means there is a strong bonding between oppositely charged **ions**. A covalent bond is formed between non-metal atoms, and each bond consists of a shared pair of **electrons**.

To understand ceramics an engineer needs to understand atomic structures.

**MATHS & SCIENCE**
Crystals are solids with regular geometric shapes, formed from regular arrangements of particles.

**MATHS & SCIENCE**
Ionic bonds form when a metal reacts with a non-metal.

### Atoms

All materials are made up from atoms. Every single object, from trees to air, is made up of the same atoms. What makes them different is the way these atoms are put together. The diversity in the properties of ceramics can be explained by consideration of their bonding and crystal structures – the way that the atoms are arranged.

**MATHS & SCIENCE**
An atom is the smallest part of a substance which can exist and still retain the properties of the substance.

Crystals exist in many different shapes and sizes. This is due to the different arrangement and bonding of the atoms. The arrangement in

**lattice** – regular repeated 3-D arrangement of atoms, ions or molecules

Close packing

space of the atoms and the way in which they are joined is called a crystal **lattice**.

All the atoms in a single element are the same size as each other.

You can imagine atoms by thinking about a box full of tennis balls. If you pack them closely in the box, the central ball will touch six others formed in a pattern around it. You can pack a second layer of close-packed balls on either the B-sites or the C-sites (but not both). A stack of layers of type ABC represents the cubic close-packed atomic structure of gold. Scientists often call this structure a **face-centered cubic** or FCC. Another form of close packing, seen in sodium atoms, is called **hexagonal close packing** or HCP. The third most common type of structure is the **body-centred cuboid** or BCC.

Of course, it would be very simple if all atoms were the same size, like tennis balls. What makes materials so different is that compounds consist of more than one element. In compounds, large atoms pack together leaving small gaps for smaller atoms to fit into.

The small atoms simply fit into the spaces that are left.

The shape of a particular crystal depends on its crystal lattice and how this lattice can be split.

It is not only the atomic structure that gives a material its properties. To understand the science more fully we need to look at a single atom.

**MATHS & SCIENCE**

**A neutron has no electrical charge.**

Atoms consist of **neutrons**, **electrons**, and **protons**.

FIG 7.1: Electrons spin around the nucleus of the atom

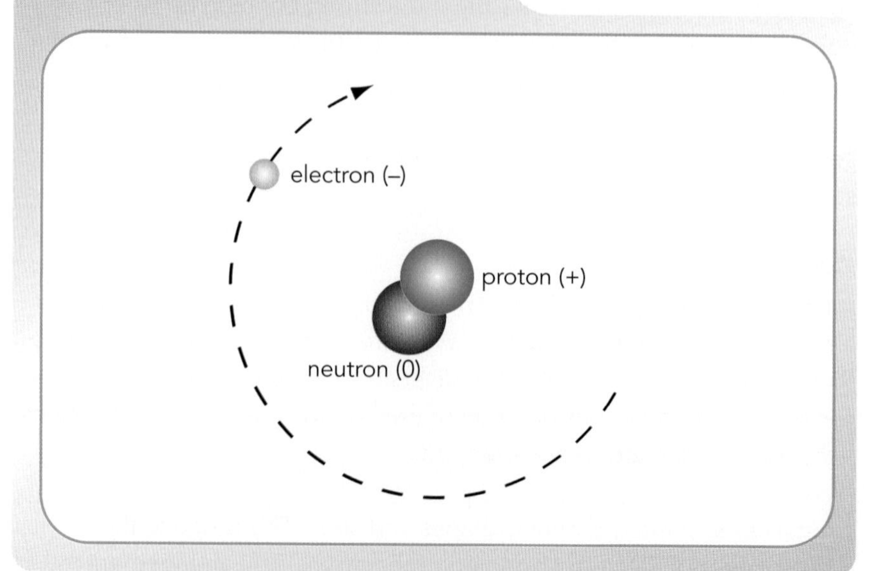

There are over 100 elements in the periodic table. The thing that makes each of those elements different is the differing number of electrons, protons and neutrons.

Scientists call the centre of the atom the **nucleus**. The protons and neutrons are always in the centre of the atom, in the nucleus. The electrons spin around the centre of the atom in areas called **orbitals**.

The electrons always have a negative charge. The protons always have a positive charge. If the charge of an entire atom is 0, that means that there are equal numbers of positive and negative pieces, an equal number of electrons and protons.

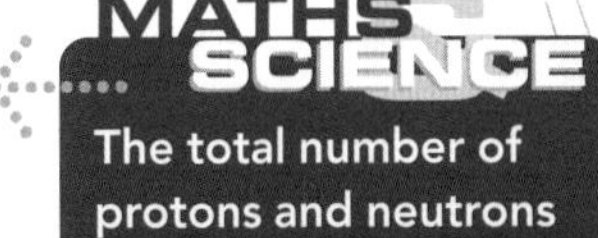


The total number of protons and neutrons in one atom of an element is given by the mass number of that element.

The atoms in a ceramic are joined together into giant structures in which electrons are shared between the bonding atoms (covalent bond), or in which electrons are transferred from one atom to another (ionic bond).

FIG 7.2: Atomic lattice of diamond

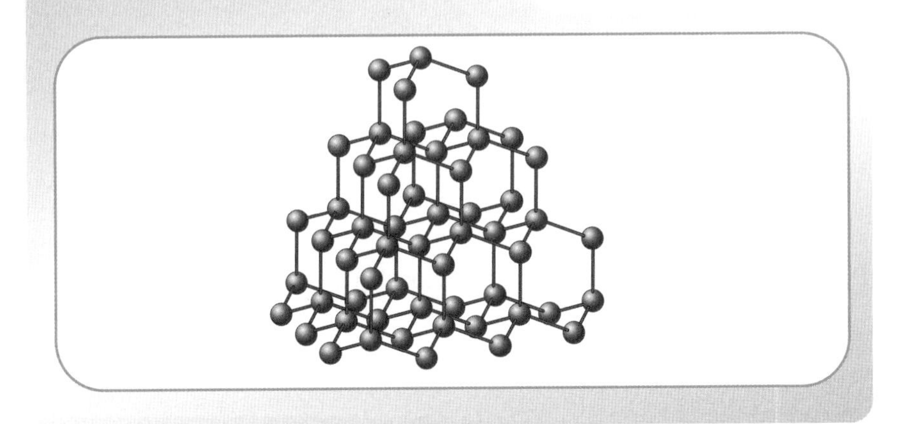

In a covalent lattice the atoms are held together by covalent bonds, forming a giant regular lattice. The structure is extremely strong because of the number of bonds.

FIG 7.3: Ionic lattice of sodium chloride

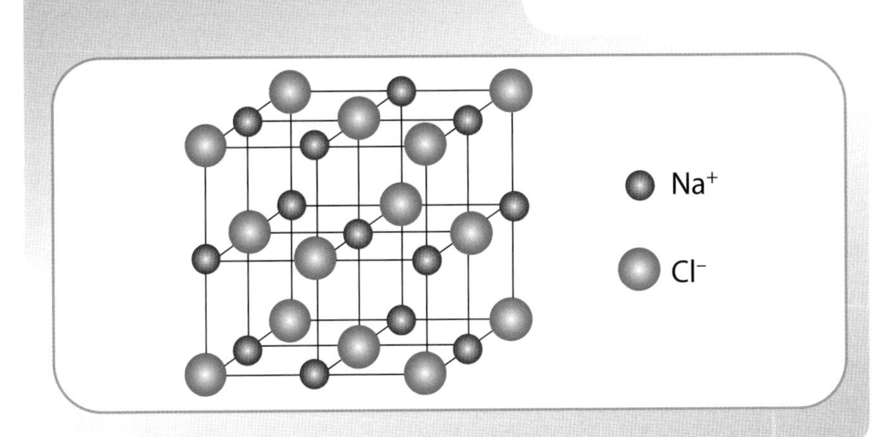

In an ionic lattice, the oppositely charged ions are arranged in a regular way to form giant ionic lattices.

Moving one atom past another would involve the breaking and remaking of these strong, directional bonds, and this would require high forces. So ceramics are strong, stiff materials.

In contrast, metals consist of regular arrays of closely packed atoms with free electrons. Specific electrons are not shared between specific atoms. They can belong to any of the atoms in the lattice and are free to move.

FIG 7.4: Metallic lattice illustrating free electrons

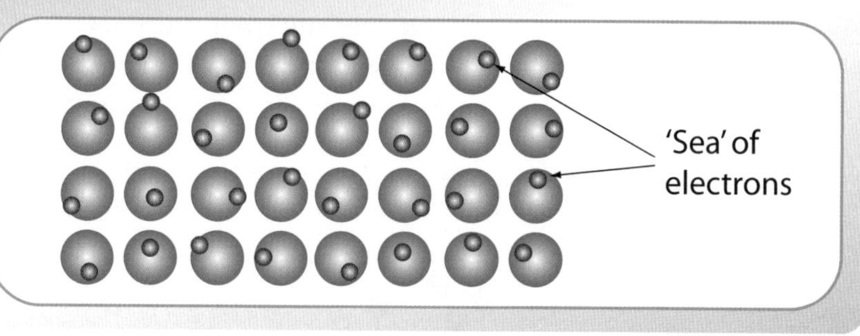


MATHS & SCIENCE

Metals are good conductors of electricity and heat because the free electrons carry charge or heat energy through the metal.

The layers of atoms can slide over one another, making metals malleable and ductile.

# Material properties

The chart below summarises the different types of material properties resulting from a particular atomic structure.

**Properties of materials**

| Structure | Mechanical | Physical |
|---|---|---|
| Atomic bonds: metallic, covalent, ionic | Strength | Density |
| Crystalline | Ductility | Melting points |
| Amorphous | Elasticity | Specific heat |
| Polymer chains | Hardness | Thermal conductivity |
| | Fatigue | Thermal expansion |
| | Toughness | Electrical conductivity |
| | Fracture | Magnetic properties |
| | | Oxidation |
| | | Corrosion |

Ceramics are often chosen because of their physical properties.

**Mechanical properties** describe the way that a material responds to forces, loads and impacts. The mechanical properties of materials were discussed in the Jaguar Cars case study.

**Thermal properties** describe the way that a material responds to heat.

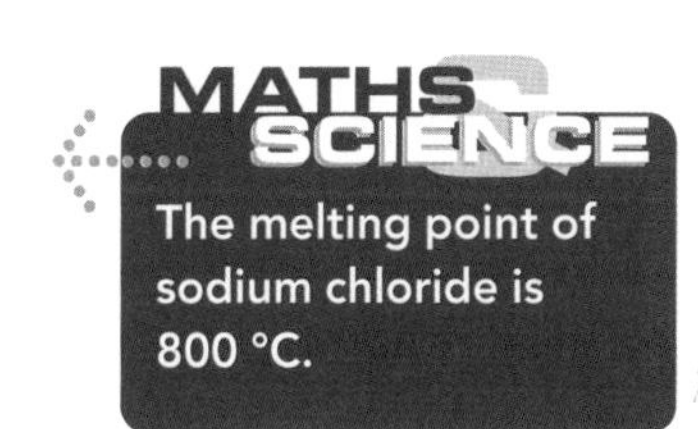


The **melting point** of a material is the temperature at which the solid changes to a liquid. For ceramic materials, due to the high strength of the bonds, or the number of bonds, a lot of energy is needed to break them. Consequently, ceramic materials have high melting points.

**Thermal conductivity** relates to the way in which heat flows through and within the material. Covalently bonded materials, like ceramics and plastics, usually have poor conductivity. Because of the large differences in thermal conductivity properties between individual materials, care needs to be taken in the selection of materials.

The **thermal expansion** of materials can have several significant effects.

Almost without exception, solids expand upon heating and contract on cooling. Remember that all materials are made up of atoms. At any temperature above **absolute zero** (-273 degrees Celsius) the atoms will be moving. Heat is energy. When a material is heated, the kinetic energy of that material increases and its atoms and molecules move about more. This means that each atom will take up more space due to its movement. This means the material will expand and get larger (T2).

**absolute zero** – the lowest temperature possible

When it is cold the kinetic energy decreases, so the atoms take up less space and the material contracts (T1).

MATHS & SCIENCE

Kinetic energy is the energy associated with movement.

FIG 7.5: Contraction and expansion

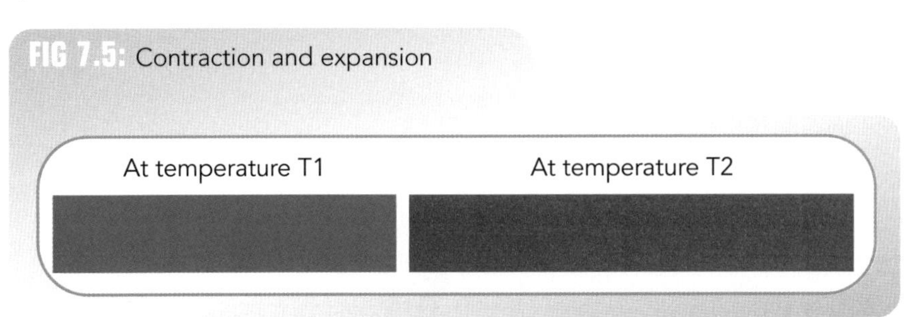


When heated, some metals expand more than others due to differences in the forces between the atoms. The linear expansion of a material can be measured by a quantity known as the **coefficient of linear expansion**.

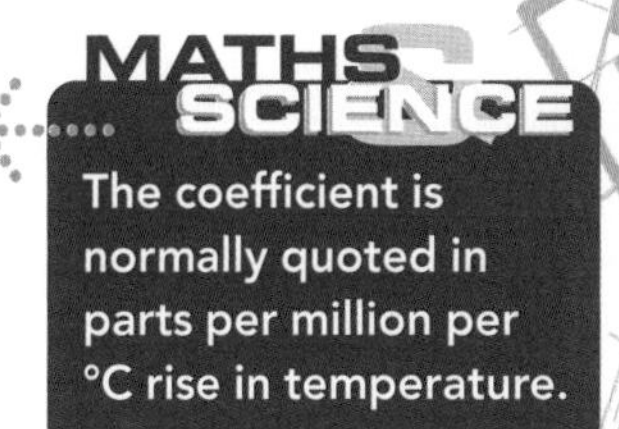


The coefficient measures the change in the length of the material per degree of temperature change.

Ceramic Seals uses a range of materials. The chart in Figure 7.6 shows the coefficient of expansion for each of the materials it uses.

In materials such as ceramics, the forces between the atoms are stronger, so it is more difficult for the atoms to move around and thus they do not expand greatly when heated.

**FIG 7.6:** Coefficient of expansion of materials used by Ceramic Seals

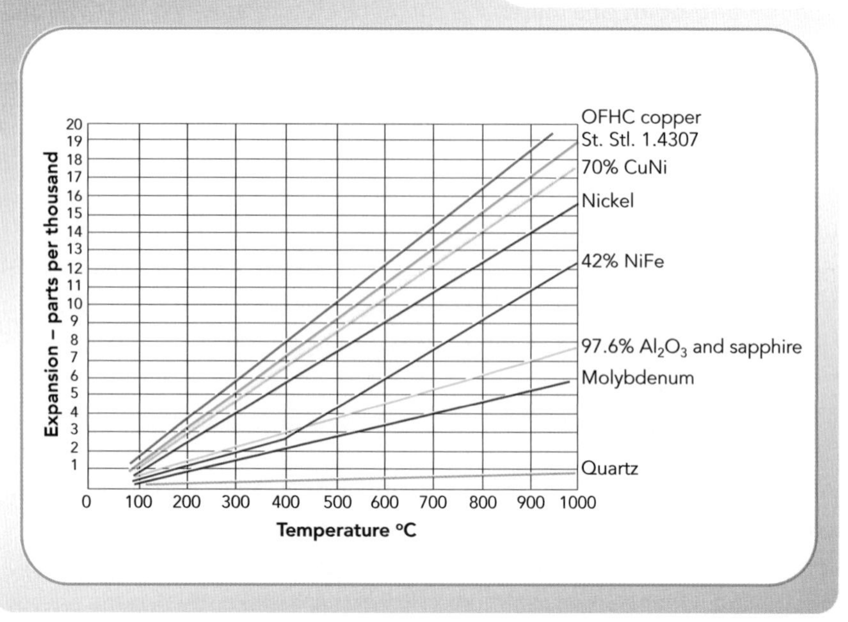


**amorphous** – a solid without a regular shape

One of the most interesting high-temperature applications of ceramic materials is their use on the space shuttle. Almost the entire exterior of the shuttle is covered with ceramic tiles. These are made from high-purity **amorphous** silica fibres.

The tiles that are exposed to the highest temperatures also have an added layer of high-emittance glass. These tiles can tolerate temperatures of up to 1480 °C.

**FIG 7.7:** The space shuttle is exposed to a range of temperatures

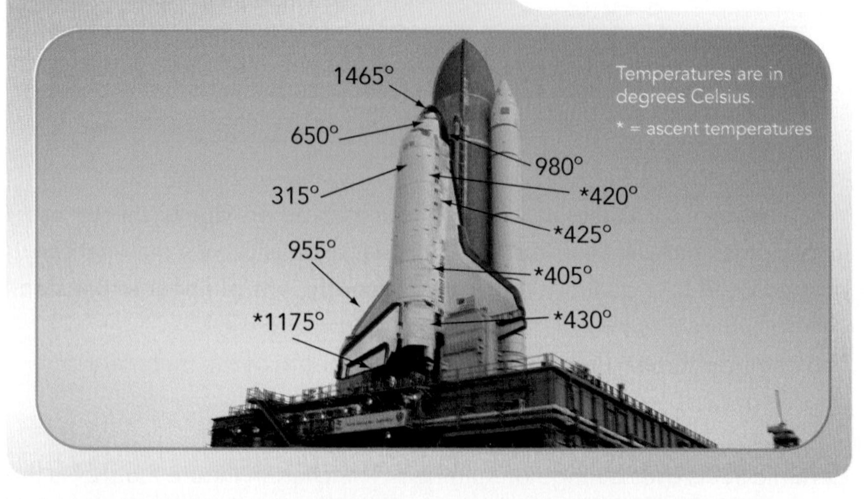


To keep it light, the body of the space shuttle is made from aluminium, which has a melting point of 660 °C. The ceramic tiles, which are only 1 cm to 7 cm thick, keep the temperature of the aluminium alloy shell of the shuttle at or below 175 °C, while the exterior temperatures can exceed 1400 °C.

The tiles cool off rapidly, which means that after prolonged exposure to high temperatures the tiles are cool enough to be held in the hand in about 10 seconds.

# Processesing of ceramics

## Raw materials

One of the most widely used materials is aluminium oxide, $Al_2O_3$.

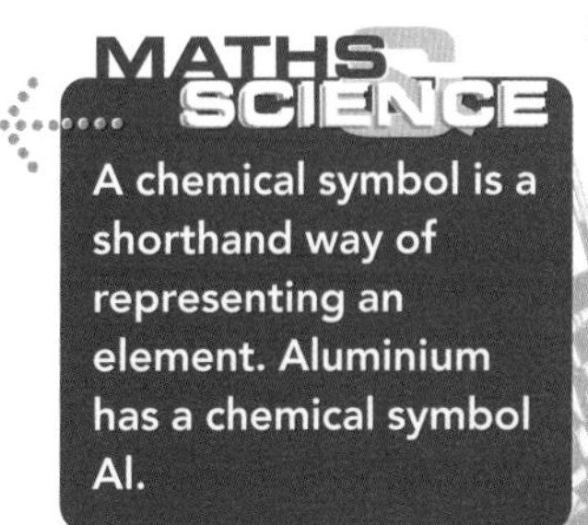


Aluminium oxide, or alumina, is used in a wide range of applications, including as insulators for spark plugs.

The raw materials from which this high-performance ceramic is made are easily available and competitively priced, resulting in good value for money.

The alumina content used by Ceramic Seals can vary with the particular application, but is normally in the range of 97.6 per cent to 99.5 per cent. This material provides:

- high electrical resistance
- corrosion resistance
- good thermal shock resistance.

It also provides consistency in these properties.

The chart below shows the technical specification for alumina.

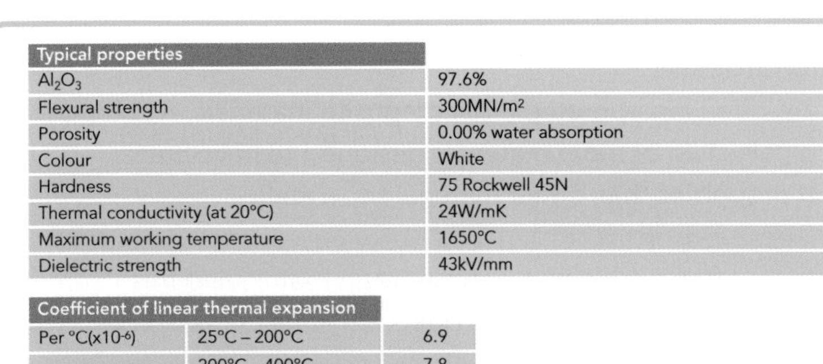

| Typical properties | |
|---|---|
| $Al_2O_3$ | 97.6% |
| Flexural strength | $300MN/m^2$ |
| Porosity | 0.00% water absorption |
| Colour | White |
| Hardness | 75 Rockwell 45N |
| Thermal conductivity (at 20°C) | 24W/mK |
| Maximum working temperature | 1650°C |
| Dielectric strength | 43kV/mm |

| Coefficient of linear thermal expansion | | |
|---|---|---|
| Per °C($x10^{-6}$) | 25°C – 200°C | 6.9 |
| | 200°C – 400°C | 7.8 |
| | 400°C – 600°C | 8.5 |
| | 600°C – 800°C | 8.8 |
| | 800°C – 1000°C | 9.0 |

| | | 25°C | 300°C | 500°C |
|---|---|---|---|---|
| Dielectric constant | 10MHz | 9.53 | 9.91 | 10.14 |
| | 1000MHz | 9.00 | - | - |
| | 8500MHz | 9.04 | 9.32 | 9.54 |
| Dissipation factor (tanδ) | 10MHz | 0.00004 | 0.00016 | 0.00052 |
| | 1000MHz | 0.00030 | - | - |
| | 8500MHz | 0.00045 | 0.00040 | 0.00072 |
| Volume resistivity | 25°C | $>10^{14}$ | | |
| Ohm-cm | 300°C | $1.0 \times 10^{12}$ | | |
| | 600°C | $2.3 \times 10^{10}$ | | |
| | 900°C | $5.0 \times 10^{8}$ | | |

The characteristics of the raw materials are very important as they affect the structure (e.g. grain size) and properties (e.g. strength) of the final component. Most of the raw materials are ground to produce a fine powder.

## Sintering

Sintering is the technique used to make an object from the powder. Sintered materials are made by converting various mixtures of powder into solids by heat and pressure.

The ISO definition of the term reads:

> The thermal treatment of a powder or compact at a temperature below the melting point of the main constituent, for the purpose of increasing its strength by bonding together of the particles.

The material is heated to below its melting point, hot enough for the particles to adhere to each other.

Ceramic sintering is not new. It is also part of the firing process used in the manufacture of traditional ceramics, such as pottery and bricks. To make any clay object durable, it must be **fired** at high temperatures in a kiln. The firing temperature is around 1000 °C. Any remaining water around the crystals evaporates. When the object cools, the liquids solidify to form glassy bridges between the crystals, gluing them together and giving the material its greatly increased strength. This process physically links the crystals because they melt into each other.

**fire** – bake objects made of clay in a kiln so that they harden
**alloy** – a mixture of two or more metals, or a metal and a non-metal

## Composites

Ceramics are often incorporated into composite materials. A composite material contains two or more materials designed to have better mechanical properties than either material separately.

Glass fibres are often added to a polymer resin matrix, and this composite is widely used in bicycles, boats and cars as glass-fibre reinforced plastic.

## Joining ceramics to other components

Ceramic Seals bonds metals to ceramics through a process known as **brazing**.

Brazing joins materials by heating them in the presence of a filler material having a melting temperature above 450 degrees Celsius, but below the melting temperature of the materials being joined.

Metallised ceramics are brazed using a selection of filler **alloys**.

# The product

## Plug and socket for vacuum chamber

In the film, you will see the manufacture of a plug and socket for a vacuum chamber.

Three pieces are shown: a metallised ceramic housing, a pin and a re-entrant.

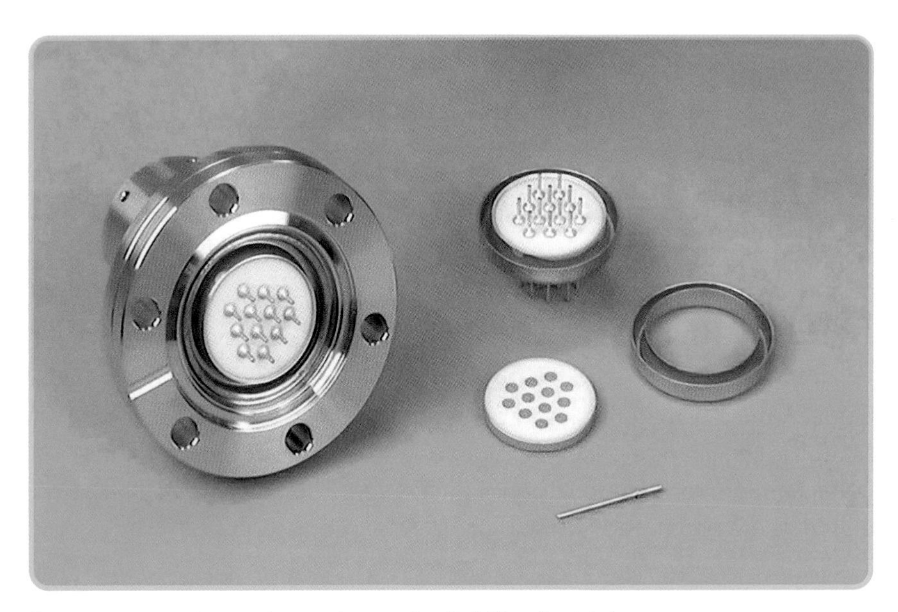

To ensure maximum **integrity** and reliability for high-vacuum applications, only certain metals can be used for the construction of this particular product.

The pin and re-entrant are made of 42 per cent nickel iron, which closely matches the thermal coefficient of expansion of the ceramic. It is important that all the different components expand at similar rates when heated.

**integrity** – the condition of being sound in construction

Nickel, chemical symbol Ni, is a silvery, metallic element with atomic number 28. It is malleable.

**MATHS & SCIENCE**

A ferromagnetic material which does not retain its magnetism after being magnetised, e.g. iron, is called soft.

**ferromagnetic** – material which is strongly magnetic

**flange** – flat surface sticking out from an object

Nickel, iron and cobalt are the only three elements known to be **ferromagnetic**. Of the three, nickel is the least magnetic. When all three ferromagnetic metals are alloyed together, an unusually strong ferromagnetic material is created. This alloy conducts heat and electricity well, but it is not as good a conductor as pure silver or copper. However, due to the high coefficient of expansion of copper, this could not be used for the pins as the joint would break apart as a result of expansion. High working temperatures would also be a problem for copper and silver.

The 42 per cent nickel-iron pin and alumina ceramic housing are brazed together to produce the sub-assembly. Brazing these materials uses temperatures of up to 850 degrees Celsius.

Ceramics are used as they are a good electrical isolator. They also withstand high temperatures. Tolerances are vital. The pin has to be brazed so it is strong enough to withstand repeated plug removal and insertion. The pins must also be vacuum-tight. The sub-assembly is welded into the stainless steel **flange**.

# The process

## Planning

### Pricing

Some of the raw materials used by Ceramic Seals are very expensive. Materials can change price overnight due to the price of metal on the metal markets and due to currency fluctuation. Therefore it is vital that materials are purchased 'just in time' (JIT) for production.

JIT systems have the following goals:

- All raw material supplies are received 'just in time' to be used.
- Product parts are made into sub-assemblies 'just in time'.
- Sub-assemblies are produced 'just in time' to be assembled into finished products.
- The company produces and delivers finished products 'just in time' to be sold.

In JIT manufacturing, parts are only produced when they are needed. This means no stockpiles. In JIT, parts are inspected as they are manufactured.

Where products require components that have been produced by outside suppliers, these also need to be delivered 'just in time' for the assembly or sub-assembly. Suppliers are often expected to deliver on a daily basis. Reliable transport systems and good, close cooperation with suppliers are vital in these situations.

## The design

Each individual part of a product has its own drawing with its own unique drawing number.

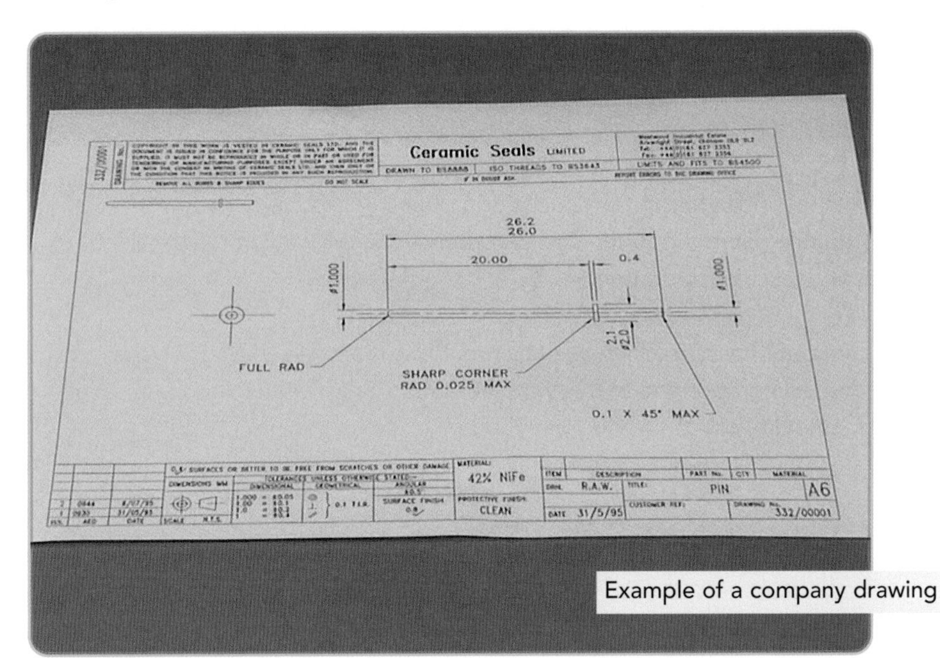


Example of a company drawing

The drawing contains all the information required to produce that part.

The specification will include:

- type of material
- tolerance to be worked to
- surface finish required.

Ceramic Seals also produces a route card for each product, containing a unique works order number.

The route card shows all the steps needed to complete production, and specifies the unique number of the drawing that is to be worked from.

These tracking sheets are part of the manufacturing process planning. Each stage of the manufacture of every part is carefully logged when

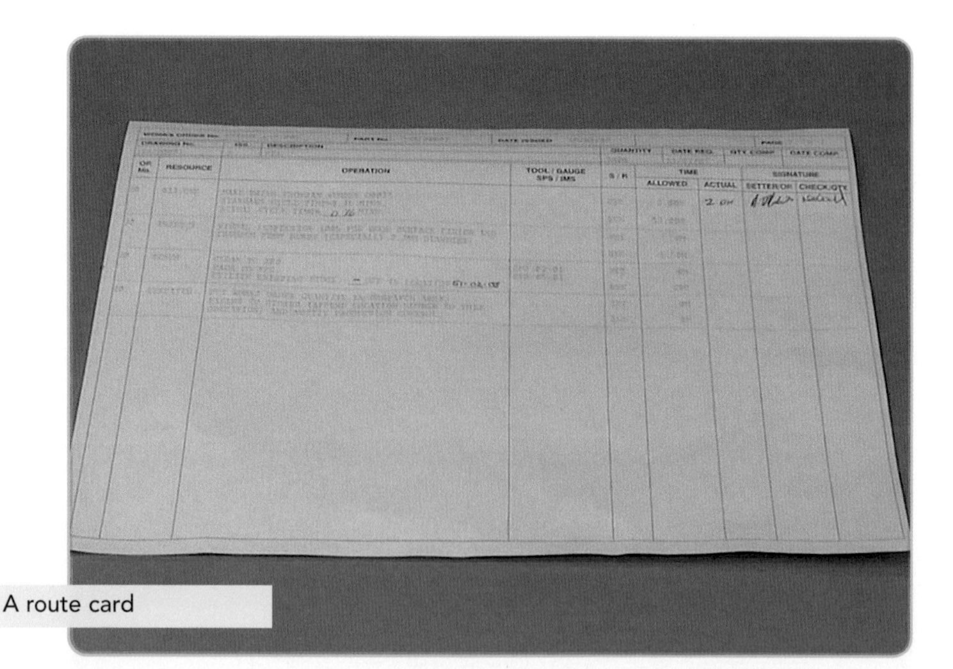

A route card

progressing through the production phase. As operators complete a step, they sign the route card and record the time taken for completing that particular operation.

The route cards track every step, from raw materials through to the finished component. In this way, Ceramic Seals is able to guarantee full traceability of the finished product.

# Production

## Equipment

Ceramic Seals use some of the very latest CNC (computerised numerical control) machines.

The main machine seen in the film is a Citizen machine. This machine has simultaneous control of eight axes.

The tool layout is extremely flexible through the wide selection of tool holders. Various configurations can give:

- up to 5 turning tools and 4 rotary tools on the vertical tool holder
- up to 3 back end machining tools
- up to 10 rotary tools on the turret (front or back end working)
- up to 40 stationary tools on the turret using multiple tool holders.

Some of the advantages of using CNC machines are as follows:

- Flexibility of operation is improved, as well as the ability to produce complex shapes with very good dimensional accuracy.

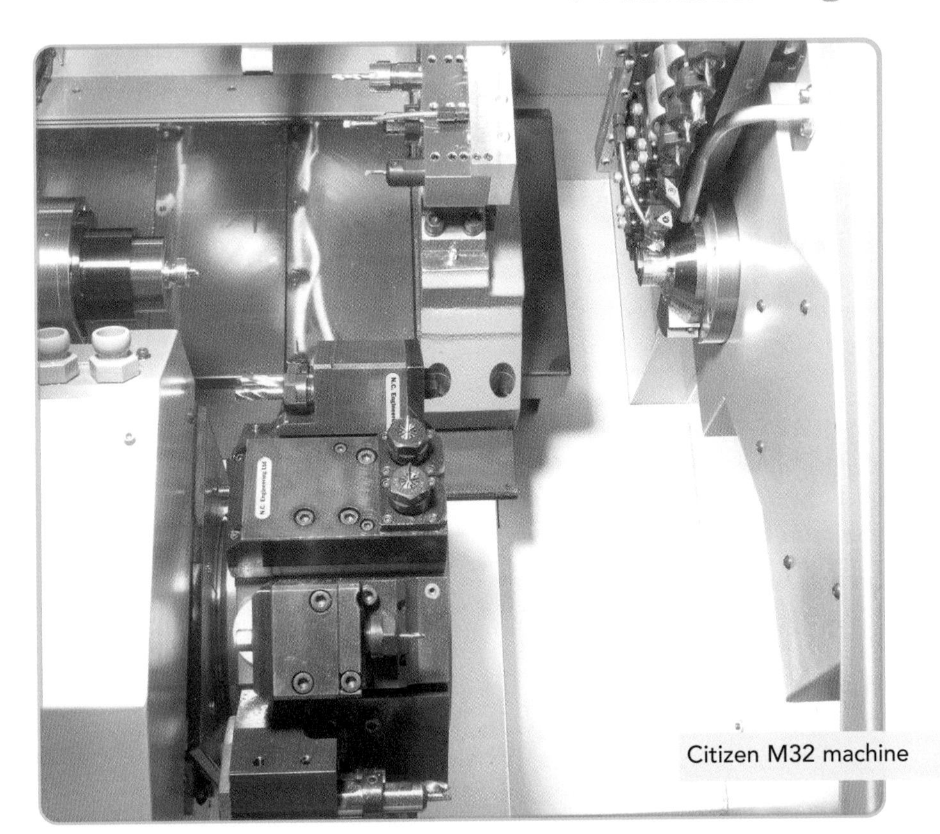
Citizen M32 machine

- There is good repeatability, as products can be produced rapidly and automatically.
- There is usually reduced scrap loss.
- There are normally high production rates and improved productivity.
- Tooling costs are often reduced because jigs and other fixtures are not required.
- More operations can be performed due to multi-tool systems.
- The machines can be run on their own without operators.

With all these advantages it is hard to see at first sight why a manufacturer would not wish to use CNC machines.

However, CNC machines have a relatively high initial cost. It is also important that operators of CNC machines are highly skilled.

## Critical point

**Tolerance** in engineering is the permissible limit of variation in a dimension or measured value, like surface finish or seal pressure. It can relate to the height, diameter or angle of the part.

Dimensional tolerance is vital on parts that have to be assembled or joined with another part.

**HEALTH & SAFETY**

Where personal protective equipment is used, it is vital to ensure that it is properly worn and sufficient for its purpose.

**NUTS & BOLTS**

To find out more about CNC machines go to www.technologystudent.com/cam/camex.htm

Dimensions and properties can vary within specified tolerance limits without significantly affecting the way the product or component works. Tolerances are specified to allow for imperfections in manufacturing parts and components without compromising performance. It is almost impossible to achieve perfection. Tool wear, measuring inaccuracy and temperature can all affect accuracy.

Tolerances can be specified as a factor or percentage of the nominal value, or more commonly as a maximum and minimum deviation from a set size.

It is good engineering practice to specify the largest possible tolerance acceptable. Closer or tighter tolerances are more difficult to achieve, and cost more to achieve in terms of machinery and scrap products that fail to meet the tolerance.

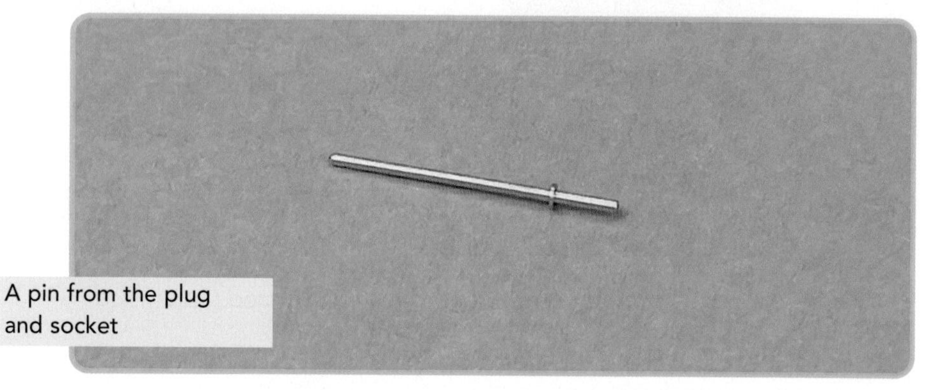

A pin from the plug and socket

Ceramic Seals works to very tight tolerances, as the components they produce work in incredibly demanding conditions. The degree of precision needed by the component parts can be one thousandth of a millimetre.

# Testing

Testing of the seals is a vital part of the process.

## Leak detection

Vacuum testing is carried out using helium as it is the smallest molecule.

Helium tests use a **mass spectrometer**. These are sensitive and accurate instruments capable of indicating extremely small leak rates.

Helium detection is a non-destructive test which is capable of quickly determining the integrity of a seal. Helium is a gas; it is very light, inert, and is present only in very small amounts in the air (approximately 4 ppm).

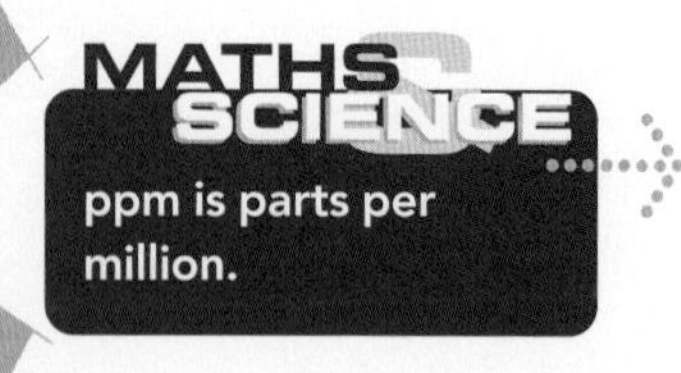


**ppm is parts per million.**

Using a helium mass spectrometer, Ceramic Seals can identify whether a product is leaking. They can even identify the location of the leak.

A helium mass spectrometer

## Tensile strength

The brazing of the pins is tested for strength, and is tested to a higher force than the joint requires.

The pins are tested using tensile strength testing, which is a form of testing to destruction.

A tensile test is probably the most fundamental type of mechanical test you can perform on a material. Tensile tests are simple, relatively inexpensive, and fully standardised. By pulling on something, you will very quickly determine how the material will react to forces being applied in tension. As the material is being pulled, you will find its strength, along with how much it will elongate before it pulls apart.

In the film you will see the pins being pulled until the seal breaks.

## Testing surface thickness

The composition and thickness of the surface coating of the ceramic are critical elements of the process.

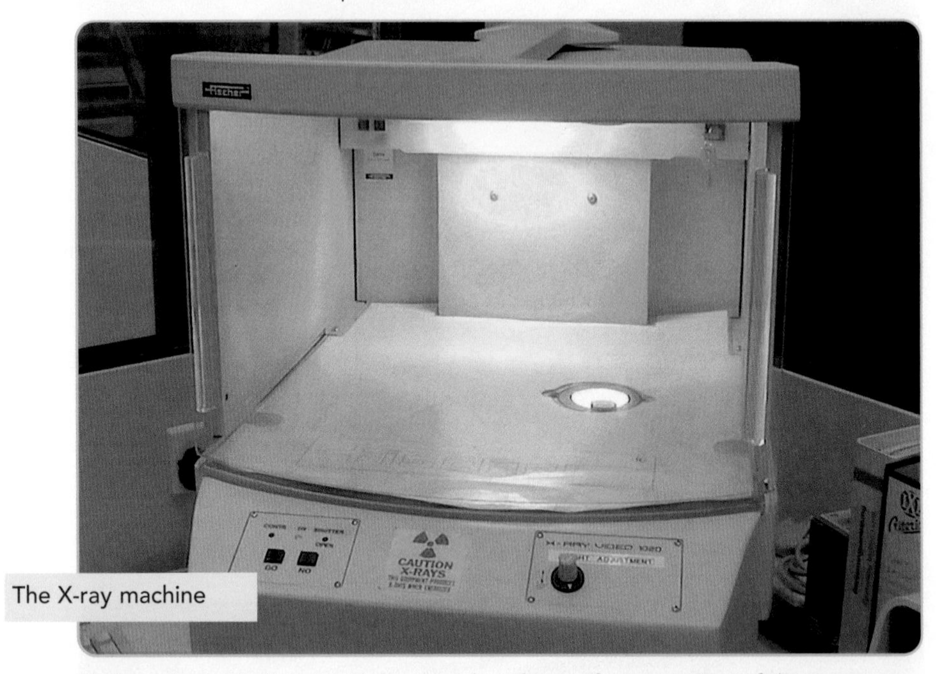

The X-ray machine

During the test, X-rays are beamed at the surface coating of the ceramic. The coating produces its own X-rays which are collected by the machine. These X-rays can then be analysed by the computer.

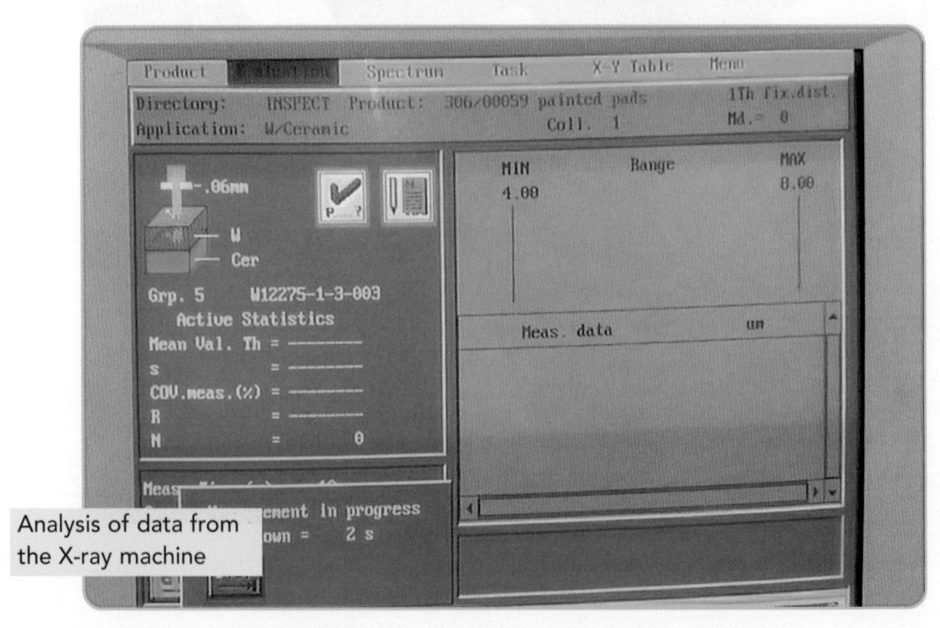

Analysis of data from the X-ray machine

The wavelength of the X-ray reveals the composition of the surface coating, i.e. what the coating is made of.

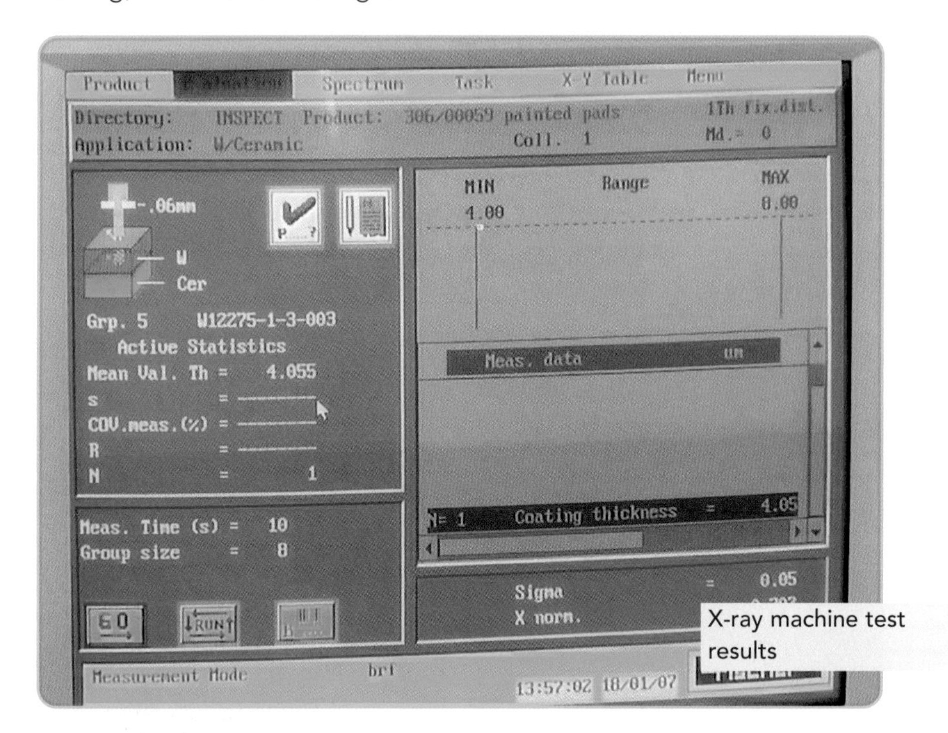

X-ray machine test results

The intensity of the beam reveals the thickness of the surface coating. The thickness must fall between two specified levels (a minimum and a maximum level) for the process to be successful.

## Coordinate measuring machine

Ceramic Seals uses a coordinate measuring machine for dimensional measuring. It is a mechanical system designed to move a measuring probe to determine the coordinates of points on the surface of a workpiece.

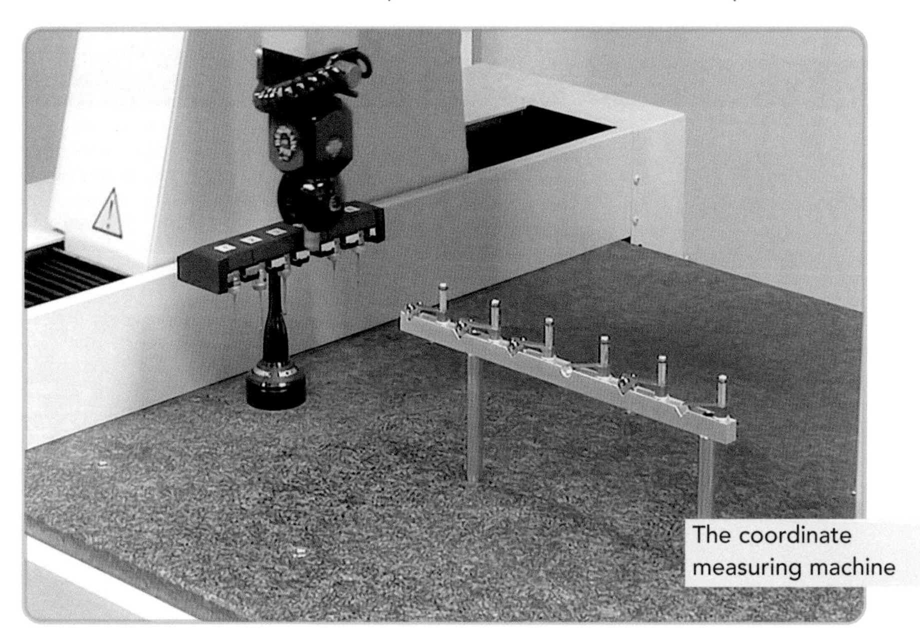
The coordinate measuring machine

The work being measured is placed on a flat table. A probe is attached to a head which is capable of various movements. The machine can record measurements very accurately. A wide range of tactile probes can be used for measuring. It is also possible to use laser and vision probes, all of which are non-tactile.

Coordinate measuring machines are very versatile and capable of recording measurements of very complex profiles to a very high resolution. The machine shown in the film can measure the tolerance of the components to three thousandths of a millimetre.

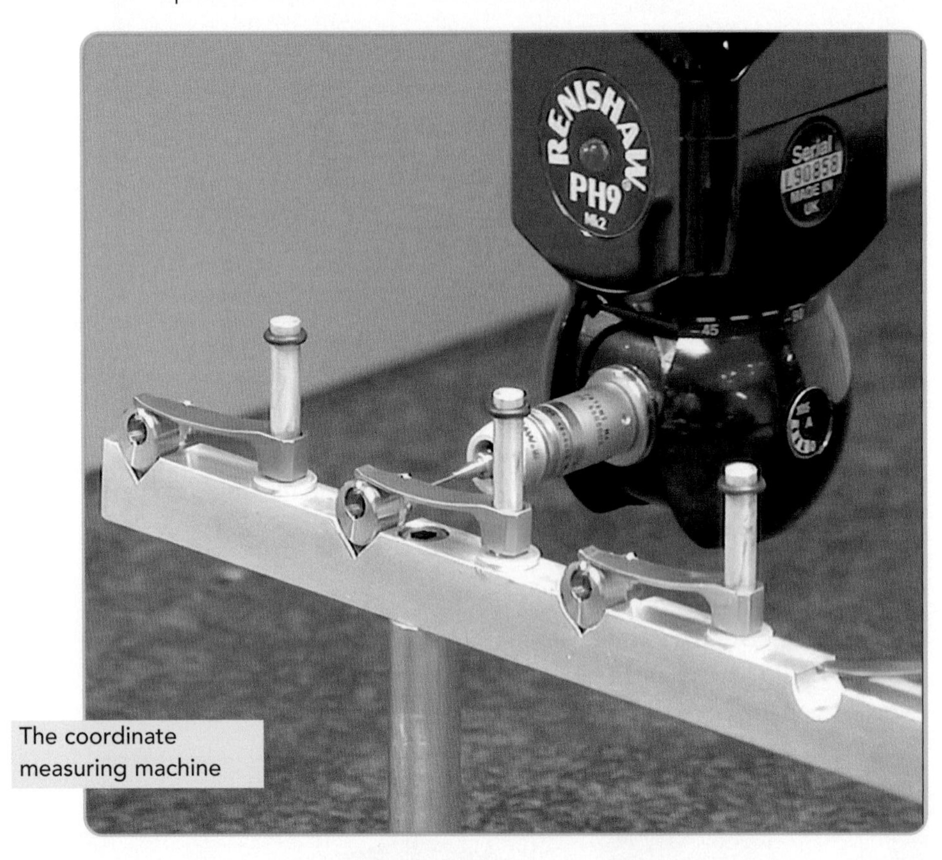


The coordinate measuring machine

# The career path

## PERSONAL PROFILE: Brett Mason

### What did you do after you left school?

I left school after completing my GCSEs and started at Oldham College as a full-time student. However, after starting at college I decided I did not want to continue in full-time education. I saw an advert in the newspaper for OTC Training – an organisation that finds apprentices for local businesses and provides practical-based training. I went for an interview and began an apprenticeship.

I began a foundation course with the intention of becoming an electrical engineer. I decided, however, to change courses, from electrical to mechanical engineering. This was because I realised that I enjoyed making things rather than simply wiring circuits.

### What about qualifications?

I have an NVQ Level 2 in welding, wiring and testing electrical circuits, fitting and using hand tools, and machining and materials. I also have Key Skills Level 2 and a City & Guilds Technical Certificate.

### What is your role at the company?

I am an apprentice. I am currently completing the practical part of my training so that I can gain my NVQ Level 3 in CNC machining. My day-to-day work involves loading the CNC machines, setting and programming them.

### Have you continued your education?

I have returned to Oldham College and attend one day a week. At college I am studying for a BTEC National Certificate in Mechanical Engineering.

### Do you enjoy college?

When I first went to college, as a full-time student, I found the work difficult. I found it hard to understand what the lecturers were talking about. I did not think that education was for me. Now that I have gone back I find it easier. Since working as an apprentice I have a better practical knowledge of the subject.

### Is working and studying difficult?

Doing two qualifications at once involves a lot of work. I have to spend one to two hours a night studying to make sure that I keep on top of it all. I have to keep my folder up to date so I write up what I have done at work each day. I also get a lot of assignments from college. But the hard work is worth it. I am getting the best of both worlds: experience and knowledge.

**NUTS & BOLTS**

**To find out more about a career in engineering go to www.learndirect-advice.co.uk/helpwithyourcareer/ and www.materials-careers.org.uk**

- have a basic understanding of global warming
- have an understanding of heat transfer and methods of reducing solar heat gain to a building
- be able to identify thermoplastics and thermosetting plastics
- have an appreciation of the stages of a tendering process
- have an appreciation of modelling techniques used to investigate the thermal environment and energy efficiency of a building
- have some ideas as to how you can become part of the building services sector.

# BUILDING SERVICES – MECHANICAL DESIGN: TROUP BYWATERS + ANDERS

## Case study overview

This case study introduces you to the building services sector and, in particular, the mechanical design of buildings.

Over the following pages you will become familiar with a real company that specialises in building services engineering.

After considering the basic principles of global warming, you will follow the company's procedures through consideration of the mechanical design of a building with specific internal temperature requirements. This will help you to appreciate thermal modelling techniques and allow you to understand the impact of surrounding buildings on energy consumption.

Finally, you will be introduced to an engineer from the company who will describe the route he took to his current position.

# The sector

## What is building services engineering?

We are all affected in one way or another by buildings. First we are born in them, then we live in them, we go to school, college and university in them, work in them, and more often than not die in them. But for most people, buildings simply go unnoticed.

Building services engineering ensures a building is fit for function. The mechanical, electrical and public health requirements of buildings are all aspects of building services.

## What is a building services engineer?

Building services engineers play an important part in all our lives. Almost every building contains some form of electrical power, heating, air conditioning, ventilation, refrigeration and plumbing. It is the job of the building services engineer to design, install and maintain these facilities.

Three types of engineer work in building services engineering:

- mechanical engineers
- electrical engineers
- public health engineers.

### Mechanical engineers

Mechanical engineers design, build and maintain the heating, ventilation and air conditioning systems. They need to understand how a building works.

### Electrical engineers

Electrical engineers design, build and maintain the lighting, power supply and specialist electrical systems, such as computer network cables.

### Public health engineers

Public health engineers design, build and maintain water supplies and waste disposal. The efficient disposal of fluid waste and organic matter is necessary in order to maintain safe public health conditions within a building.

# The company

## Background

Troup Bywaters and Anders (TB+A) is an **equity partnership** which was formed in 1958.

A total of 10 people own the company, contributing money to the business whilst also working on a day-to-day basis as part of the company. All projects undertaken by TB+A are led by one of the company's partners.

### Consulting building services engineers

TB+A provides a building services design consultancy for new and refurbished buildings across a range of markets, including commercial office, health care, residential and educational sectors. Each project is dealt with individually, which ensures ideal solutions are developed for the specific circumstances of each design.

Buildings that TB+A have worked on include the New Coventry Hospital, and The Grove Hotel, Hertfordshire.

## Application of technology

TB+A continue to invest in developing their **IT infrastructure**. Their **PC** to staff ratio is 1:1.

They have undertaken many projects in the past few years that have required document issue and document control to be fully electronic. This has entailed the use of a variety of systems, including Bovis Lend Lease 'Humming Bird', Sun Microsystems 'C3', the internet provider 'Bidcom', and design-team-hosted **FTP**s.

To ensure speedy transition of documents, the company utilises dedicated **ISDN** lines. Documents can be issued via internet transmission or email. Each system has dedicated **routers** and ISDN lines.

In terms of computer-aided design, they have a full range of packages, which include:

- design calculations – Apache, Hevacomp, Amtech Ltd
- thermal modelling and computational flow dynamics – IES & TAS
- infiltration analysis – Breeze
- computer-aided design – Autocad 2000 & 2002.

**equity partnership** – a limited partnership arrangement for providing start-up capital
**IT infrastructure** – information technology infrastructure: the basic systems and services that an organisation uses in order to work effectively
**PC** – personal computer
**FTP** – file transfer protocol
**ISDN** – Integrated Services Digital Network
**router** – device which forwards data packets to the appropriate parts of a computer network

# Modern-day concerns

Assessments generally suggest that the earth's climate has warmed over the past century, and that human activity affecting the atmosphere is an important contributing, if not driving, factor.

The building sector accounts for over 40 per cent of Europe's energy requirements. It offers the largest single potential for energy efficiency.

**Most fuels used today are fossil fuels, which were formed from the remains of prehistoric animal and plant life.**

**deforestation** – clearing an area of forest or trees

**methodology** – a system of ways of doing something

NUTS & BOLTS

**To find out more about TB+A go to www.tbanda.co.uk**

## The impact of human activities

Burning **fossil fuels** releases the carbon dioxide stored millions of years ago. We use fossil fuels to run vehicles, heat homes and businesses, and power factories. **Deforestation** releases the carbon stored in trees and also results in less carbon dioxide being removed from the atmosphere.

## Energy performance of buildings

The European Commission's Action Plan on Energy Efficiency (2000) indicated the need for specific measures in the building sector. In response, the European Commission published the EU Directive on the Energy Performance of Buildings. Measures within European law include:

- **methodology** for calculating the energy performance of buildings
- application of performance standards on new and existing buildings
- certification schemes for all buildings
- regular inspection and assessment of boilers/heating and cooling installations.

## The design challenge

We cannot live without using energy, but we can design our lifestyles and buildings to be as efficient as possible. TB+A pride themselves on their tradition of energy-efficient design.

# The product

## Introduction

This case study will focus on mechanical design engineering. As the project discussed in the film focused on the requirements for energy performance, we will begin by considering the issue of global warming, and energy-efficient systems available today.

**attributable** – caused by
**infrared** – electromagnetic waves most commonly produced by hot objects
**ozone** – poisonous, bluish gas made of molecules which contain three oxygen atoms

## Global warming

The earth could be getting warmer on its own, but many of the world's leading climate scientists think that the things people do are contributing to the warming effect.

The Intergovernmental Panel on Climate Change announced in 2001 that:

> most of the warming observed over the last 50 years is likely to be **attributable** to human activities.

The term '**greenhouse effect**' is commonly used to describe the increase in the earth's average temperature that has been recorded over the past 100 years.

**Greenhouse gases** allow incoming solar radiation to pass through the earth's atmosphere, but prevent most of the outgoing **infrared** radiation from the surface and lower atmosphere from escaping into outer space.

Any gas that absorbs infrared radiation in the atmosphere is a greenhouse gas. Greenhouse gases are generally divided into two categories:

- the four principal greenhouse gases – carbon dioxide, nitrous oxide, methane and water vapour
- other gases – halocarbons and **ozone**.

Carbon dioxide is by far the most important of the greenhouse gases as it accounts for the largest proportion.

The best known gases in the halocarbons group are CFCs (chlorofluorocarbons), HCFCs (hydrochlorofluorocarbons) and HFCs (hydrofluorocarbons). While the concentration of halocarbons is much lower than those of the other greenhouse gases, the warming effect that they produce ranges from 3,000 to 13,000 times that of carbon dioxide.

**MATHS & SCIENCE**

**The trace gases which regulate the climate constitute 1% of the earth's atmosphere.**

**MATHS & SCIENCE**

**Carbon dioxide, nitrous oxide, methane and water vapour are all naturally occurring gases.**

**MATHS & SCIENCE**

**The fluorocarbons are man-made gases. Once they are in the atmosphere, they resist breakdown and don't disappear for many decades.**

Current life on earth would not exist without the '**natural greenhouse effect**'. The earth receives its warmth from the sun. On its way to the earth's surface, most of the heat energy passes through the earth's atmosphere, while a smaller proportion is reflected back into space.

The energy warms the earth's surface, and as the temperature increases, the earth radiates heat energy back into the atmosphere. As this energy has a different wavelength from that coming from the sun, some is absorbed by gases (previously identified) in the atmosphere.

Once these gases absorb energy, the gas particles begin to vibrate and they radiate energy in all directions, including approximately 30 per cent of it back towards the earth.

Excessive use of energy, landfill and the use of fossil fuels have accelerated the process. An increase in greenhouse gases in the atmosphere enhances the atmosphere's ability to trap heat, which leads to an increase in the average surface temperature of the earth.

A warmer earth leads to changes in rainfall patterns, a rise in sea level, and a wide range of impacts on plants, wildlife and humans.

# Energy use

## Commercial buildings

Energy use is an important element in designing building spaces for work and leisure. Building services such as heating, lighting, ventilation and cooling all use energy. The pie chart below shows the energy use of a typical building.

**MATHS & SCIENCE**

The composition of the pie chart adds up to 100%.

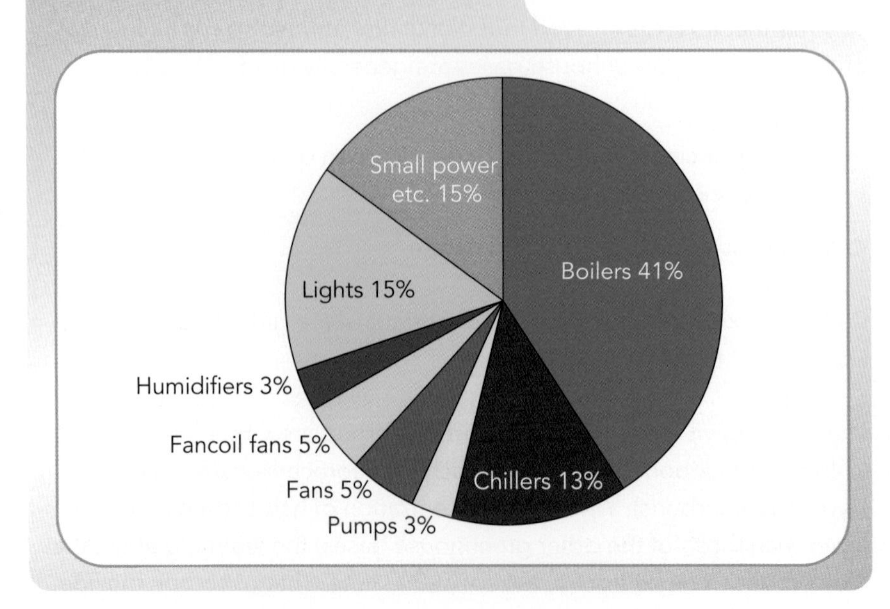


**FIG 8.1:** Energy use of an average building

The chart actually shows the annual energy use measured at the building. If we consider the same building based on fossil fuel energy requirement, we can see that the chart changes significantly.

**FIG 8.2:** Fossil fuel energy requirements of an average building

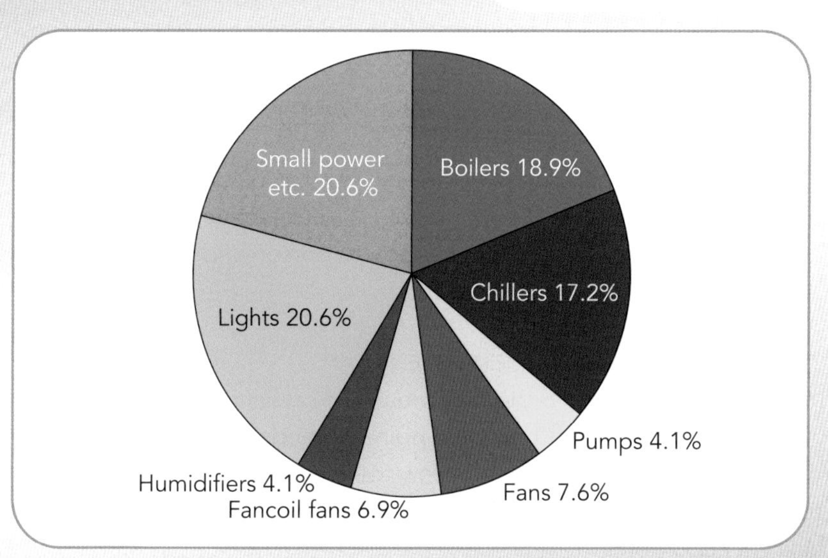


This is because electrical energy requires three times as much fossil fuel per useful kilowatt as the natural gas used in the boilers. Whilst it is true that electrical power stations are changing, presently – according to energy statistics – electricity produces three times as much carbon dioxide emissions as natural gas.

## Domestic buildings

Energy – gas and electricity – used in the home is responsible for 25 per cent of the UK's carbon dioxide emissions.

As a result of the government's energy strategy, domestic technological improvements are being **pioneered**.

- Compact fluorescent lamps, now in common use, are 70 per cent more efficient than **incandescent lamps**.
- Domestic appliances account for 47 per cent of total domestic electricity consumption, according to the Energy Saving Trust. All domestic appliances are now efficiency rated.
- The typical level of insulation in walls has increased.
- Typical insulation levels used in ceilings and attics have improved.
- The use of high-efficiency low-emissivity (low-E) coated double-glazed windows is growing.
- The use of insulated glass has increased from nearly 68 to 87 per cent.

**pioneered** – developed

**incandescent lamp** – contains a filament which glows white-hot when heated by a current passed through it

**MATHS & SCIENCE**

**A kilowatt is 1000 watts.**

All domestic appliances have an efficiency rating

# The building envelope

The **building envelope** is the **interface** between the inside (interior) of a building and the outside (exterior) environment. The envelope separates the living and working environment from the outside environment.

**interface** – place where two things come together and affect each other
**subjective** – based on personal beliefs or feelings, rather than based on facts

It is designed to provide protection from the elements (wind, rain, snow, heat and cold) and to control the transmission of cold, heat, moisture and sunlight to provide the necessary comfort for people inside the building.

In basic terms, the building envelope needs to provide the 'right' temperature for comfort. This, however, is a **subjective** condition. The 'right' temperature will vary from person to person and from time to time, i.e. daily and seasonal change.

## Ways of heat transfer

To be able to understand the behaviour of a building in response to both climate and climatic changes, the engineer needs to understand heat transfer. This understanding will allow the engineer to predict the change in the internal environment of any building envelope due to changes in weather conditions.

**MATHS & SCIENCE**

**Heated electrons gain kinetic energy and move out fast in all directions.**

The flow of heat through a material takes place by **conduction** from the warm to the cold side. In good conductors the energy transfer is rapid, occurring mainly by the movement of free electrons.

**Convection** is the way in which heat energy is transferred in liquids and gases. If a liquid or gas is heated, it expands, becomes less dense and rises. A good example of this is radiators. Hot air rises and heats the upper part of the room.

**Radiation** is the way in which heat energy is transferred from a hotter to a cooler place without a medium taking any part in the process. Radiation can occur through a vacuum. Radiation is often used to refer to the heat energy itself. This takes the form of electromagnetic waves. When these waves fall on an object, some of their energy is absorbed, increasing the object's internal energy and, as such, its temperature. Refer to the lighting design case study for more detail on electromagnetic waves and absorption.

## Maintaining a comfortable environment

A building has heat exchange processes with the surrounding environment. **Thermal equilibrium** will exist if the heat gain by a building equals the heat loss of a building.

If heat gain is greater than heat loss, the temperature in the building will increase. If heat loss is greater than heat gain, the building will be getting colder.

Sources of heat gain may include:

- conduction of heat through walls or roof assemblies
- solar radiation on glazing
- internal heat gain from the occupants within the building and the activities they are doing.

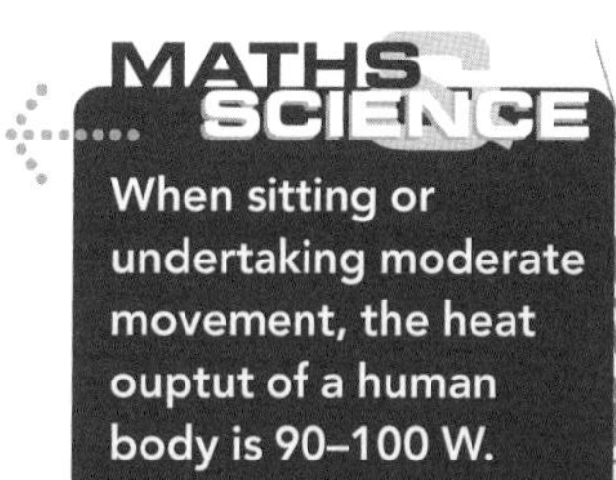


The energy released by people, machines, lighting and other sources which are not part of the heating system can have a major affect on the indoor climate, and need to be controlled.

In buildings such as office buildings and retail stores, much of the overheating problem in the warm weather can be caused by heat produced by equipment, or by a high level of artifical lighting. Refer to the case study on lighting design for more details.

# What is the building made of?

The building fabric is a critical component of any building. It protects the building's occupants and helps control the indoor environment.

The fabric of the building controls the flow of energy between the interior and exterior of the building. The roof, floor, walls, windows and doors make up the building's fabric.

Technological advancements in materials and processes are ensuring greater energy efficiency for each of these structural elements.

## Roof

The building's roof presents a large surface area exposed to year-round direct sunlight. The heat available from this source is welcome during the winter, but summertime heat gains increase the use of air conditioning and, thus, energy use.

New 'smart' roofing materials are able to absorb solar energy when the outdoor temperature is cool, and reflect solar energy when the outdoor temperature is warm.

## Floor

Energy savings from insulating the floor are significantly smaller than those of the other building elements. This is because the temperature of the ground is higher than the external temperature.

## Walls

The use of materials with high **thermal storage**, or thermal mass, can **modulate** heat gains by a building. Concrete and brick are both high

**modulate** – change something to make it more suitable for its situation

thermal storage materials. They act as storage for both heat and cold as they heat up and cool down relatively slowly.

External insulation of brickwork can extend the lifetime of a building by protecting its shell from the weather conditions.

**Thermal insulation** is the use of a material with low conductance to reduce the energy flow across another material. The insulation acts to slow and/or reduce the flow of heat. The material used must have a high resistance. Materials such as wood and most liquids and gases are insulators, i.e. they are bad conductors.

**MATHS & SCIENCE**

**Insulators do not have free electrons. Instead, when heated the hot atoms vibrate but only collide with neighbouring atoms.**

When using core insulation within a wall, the thermal insulation is protected from external influences. However, there is a possibility of vapour condensation.

### Windows

Solar heat gains can be controlled by the sizing and positioning of windows. The larger the window opening, the greater is the potential for solar heat gain and the danger of overheating. To reduce heat gains, reflecting or absorbing windows can be chosen.

New types of double glazing, which turn **opaque** at high temperatures, can be used to prevent buildings from overheating in the summer. Many glass buildings have blinds or air conditioning installed to keep their interiors comfortable in hot weather.

**opaque** – not able to be seen through

**crystalline** – having the structure and form of a crystal

The insulating gap in some new types of double glazing is filled with a thermo-optical polymer which has optical properties that change with temperature. On cool days, the polymer exists in its **crystalline** form and transmits light. However, on hot days, the polymer undergoes a phase transition into a melted state that reflects more light and infrared radiation than it transmits.

**MATHS & SCIENCE**

**Polymers are substances that consist of many small molecules bonded together in a repeating sequence. Rubber is a polymer that occurs naturally.**

## Materials

Many of the new technologies employed in buildings are as the result of developments in materials, enabling new and more efficient construction methods. Alongside developments in traditional construction materials, such as ceramics, metals and timber, developments in plastics have revolutionised the building industry.

Plastics are a very important material in the building sector. In fact, roughly 20 per cent of all the plastics used in Europe are used for products in the building industry. Plastics are used for pipes, windows and floor coverings, and as foam for insulation.

What makes plastic an ideal material for building is the fact that it can be strong, weather-resistant, heat-resistant and flexible. Plastics are also very light and they require very little maintenance. Plastics make great insulators and sealers, which helps improve the energy efficiency of buildings.

## What is plastic?

The name plastic is a little misleading. Plasticine and chewing gum are examples of materials which are truly plastic. They are easily deformed by tensile, compressive and shearing forces, and retain whatever shape they are moulded into.

In actual fact, plastics are **synthetic** or **man-made polymers**. These polymers are prepared in the laboratory or in industry.

Plastics are made from chemicals derived from petroleum, and are usually durable, light solids that are thermal and electrical insulators. However, plastics are often not biodegradable, and give off poisonous fumes when burned.

**MATHS & SCIENCE**

**Petroleum is formed over millions of years by the decomposition of animals and plants under pressure.**

There are two types of plastic – **thermoplastics** and **thermosetting plastics**.

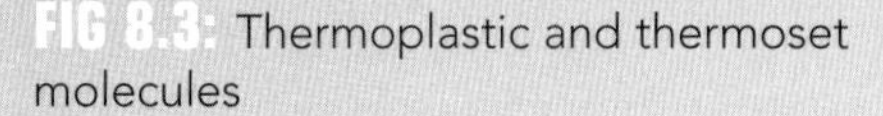

**FIG 8.3:** Thermoplastic and thermoset molecules

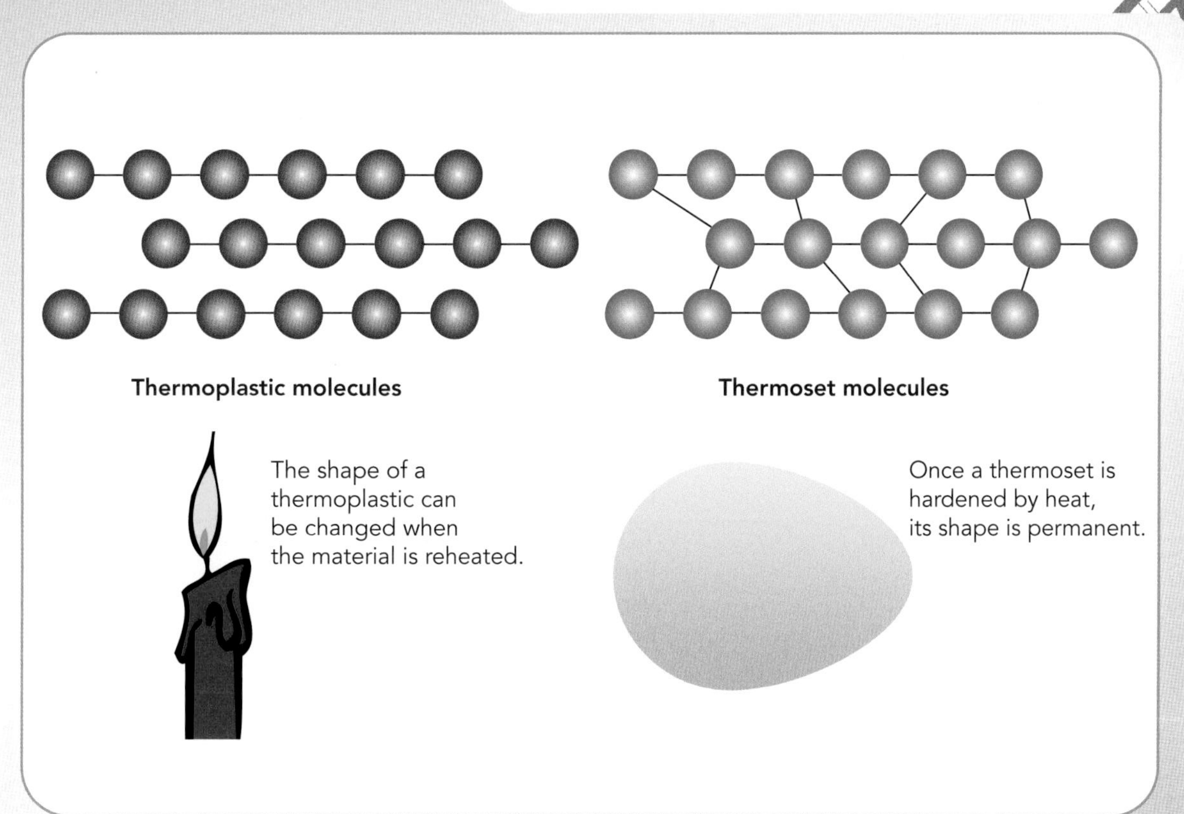

Thermoplastics are those which, once formed, can be heated and re-formed over and over again. This property allows for easy processing and facilitates recycling.

Thermosetting plastics, or thermosets, cannot be re-formed or remoulded. Once these polymers are formed in a particular shape there is no going back.

## Thermoplastics

| Polymer | Common name | Properties | Use in buildings |
|---|---|---|---|
| Polythene | Low-density polythene | Tough<br>Flexible<br>Solvent-resistant<br>Degrades if exposed to light | Pipes<br>Cable and wire insulation |
| Polythene | High-density polythene | Very tough<br>Stiff<br>High tensile strength | Pipes<br>Mouldings |
| Polypropene | Polypropene | Hard<br>High melting point<br>High strength | Tubes<br>Pipes<br>Electronic components |
| Polychloroethene or polyvinyl chloride | PVC | Can be tough and hard or soft and flexible<br>Solvent-resistant<br>Hardens over time | Window frames<br>Pipes<br>Guttering<br>Wire insulation |
| Polyphenylethene | Polystyrene | Tough<br>Hard<br>Can be made into foam<br>Degrades if exposed to solvents | Foam mouldings<br>Solid mouldings |
| Methyl-2-methylpropenoate | Perspex | Strong<br>Rigid<br>Transparent<br>Easily scratched<br>Easily softened<br>Degrades if exposed to solvents | Roof lights<br>Light fittings<br>Protective shields<br>Windows |
| Polytetrafluoroethene | PTFE | Tough<br>Flexible<br>Heat-resistant<br>Solvent-resistant<br>Low friction | Tape<br>Coatings |
| Polyamide | Nylon | Tough<br>Flexible<br>Very strong<br>Solvent-resistant<br>Absorbs water<br>Deteriorates outside | Textiles<br>Brushes |
| Thermoplastic polyester | Terylene | Strong<br>Flexible<br>Solvent-resistant | Electrical insulation tape |

## Thermosetting plastics

| Polymer | Common name | Properties | Use in buildings |
|---|---|---|---|
| Phenolic resins | Bakelite | Hard<br>Heat-resistant<br>Solvent-resistant<br>Electrical insulator<br>Machinable<br>Black or brown | Electrical fittings<br>Laminate |
| Urea-methanal resin | Formica | Hard<br>Heat-resistant<br>Solvent-resistant<br>Electrical insulator<br>Machinable<br>Transparent<br>Can be many colours | Electrical fittings<br>Toilet seats<br>Laminates |
| Methanal-melamine resins | Melamine | Very hard<br>Smooth finish<br>Very good heat resistance<br>Solvent-resistant<br>Electrical insulator<br>Machinable<br>Transparent<br>Can be many colours | Handles<br>Control knobs<br>Electrical equipment |
| Epoxy resins | Epoxy resin | Strong<br>Tough<br>Good electrical resistance<br>Good adhesive | Flooring material<br>Laminates<br>Adhesives |
| Polyester resin | Polyester resin | Strong<br>Tough<br>Wear-, heat- and water-resistant | Laminates<br>Panels |

# Heating, ventilation and air conditioning systems

The main function of heating, ventilation and air conditioning (HVAC) systems is to provide healthy and comfortable interior conditions for occupants.

Ventilation is used to remove unpleasant smells and excessive moisture, introduce outside air, and keep the interior building air circulating.

In a mechanical ventilation system, the supply air and the exhaust air are transported mechanically.

**FIG 8.4:** A typical HVAC system

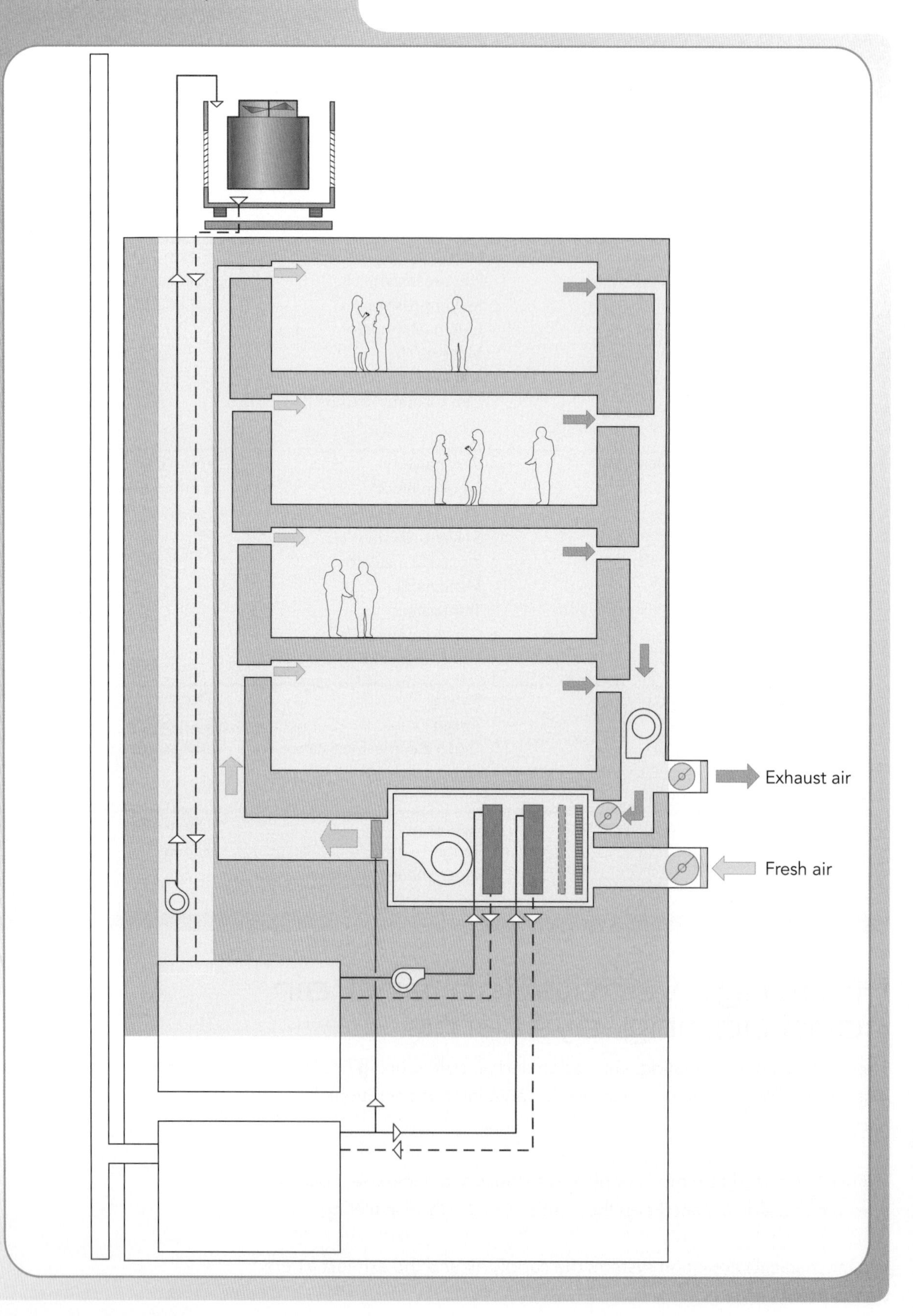

The cleanliness of the recirculated air has to be maintained by the use of filters.

HVAC systems use ventilation air ducts installed throughout a building that supply conditioned air to a room through rectangular or round outlet vents, called diffusers.

Air conditioning is the process of treating air so as to control its temperature, humidity, cleanliness and distribution. In air-conditioning processes heat is added to or extracted from the air to produce heating or cooling.

Air-conditioned buildings often have sealed windows, because open windows would disrupt the attempts of the HVAC system to maintain constant indoor air conditions.

Well designed, efficient systems do this with minimal non-renewable energy use and little or no air and water pollutant emissions. Cooling equipment that avoids chlorofluorocarbons and hydrochlorofluorocarbons (CFCs and HCFCs) eliminates a major cause of damage to the ozone layer.

However, even the best HVAC equipment and systems cannot compensate for a building design with high cooling and heating needs. The greatest opportunities to conserve non-renewable energy are through architectural design that controls solar gain while taking advantage of **passive** heating, daylighting, natural ventilation and cooling opportunities.

The critical factors in a mechanical system's energy consumption – and capital cost – are reducing the cooling and heating loads that they must handle.

HEALTH & SAFETY

**Many of the world's nations agreed to control the use of CFCs in 1987 when they signed the Montreal Protocol on Substances that Deplete the Ozone Layer.**

**passive** – not acting to influence or change a situation

# The process

## Planning

### Winning the contract

The real-life project discussed in the film is St Bartholomew's Hospital. The company won the contract in a competition. Building contracts are often offered in this way. This type of competition is not like the competitions the general public can enter. Companies have to submit their design ideas, and the best ideas are selected for a presentation. A wide range of presentation techniques are needed to win the contract, including drawings, models and reports.

The process is known as **tendering**. A tender is an offer to provide a service or a product, or to carry out work in a contract. There are many types of tendering.

**Selective tendering** is a form of competitve tendering in which only selected tenderers are invited to submit prices. Contrast this with a **negotiated tender**, which is a price for providing goods and services from a single supplier, i.e. a single contractor.

Before a bid can be won and any work can start, the company has to undertake a large amount of work.

### Brief

The key to a successfully designed building is to fully understand, at an early stage, the client's expectations of the end product. The project is defined in a detailed brief, which TB+A would prepare, in conjunction with other team members, following user group/client interviews.

The brief would typically comprise:

- room data sheets for all rooms
- concept design policy documents for all major systems.

Prior to progressing with design, the brief, together with the associated cost plan, would be signed off by the client and design team.

### Feasibility stage

A **feasibility study** develops the requirements for a project and provides an understanding of whether the project is possible, practical or viable.

Working through the feasibility stage, the company would address, as a minimum, the following key issues:

- effect on and integration with building finishes and features (plant space, **risers**, etc.)
- setting of energy targets, and options to meet or improve upon these targets
- statutory and local authority requirements
- future flexibility
- **CDM** regulations and maintenance issues.

The above key issues would be closely monitored throughout all stages of design, up to the final point of tender. At this point, the goal would be to produce (as far as practicable) a fully designed engineering services package.

## Quality assurance

Prior to tender, a quality assurance review would be held with all TB+A design team members to evaluate all project decisions, coordination of the project and solutions. The aim of the quality assurance review is to highlight and **resolve** any areas of possible risks, prior to contractual agreement.

## Tendering for the contract

During the tender period, any questions or issues arising from the documentation would be dealt with and clarified via the contract administrator.

## When successful

Acceptance, in engineering, is the consent to receive, e.g. assenting to an offer that, together with the tender, forms a contract.

When the project reaches construction stage, the company would review the contractor's working drawings and schedules with the design team. This is to ensure compliance with design intent, and also to review **procurement** progress.

Of course, with such a large contract the customer could change its mind. TB+A fully recognise that control of change at all stages of the project is critical to maintaining budgets and programme. Change control deals with the routine modifications and corrective actions needed to ensure that the project is maintained according to plan. Control of change is particularly important during the construction stage and, as such, it is important to embrace a fully coordinated change control system prior to commencement of work on site.

Engineers would attend site to see that the quality of workmanship generally adheres to the standard expected by the client and themselves.

**HEALTH & SAFETY**

**The Construction (Design and Management) Regulations 2007 are aimed at improving the overall management and coordination of health and safety and welfare throughout all stages of a construction project.**

**riser** – vertical pipe for the upward flow of liquid or gas
**resolve** – to solve or end a problem
**procurement** – complete process starting from identifying and specifying need and finishing with the making of a purchase

Progress of works will be monitored and any deficiencies resolved with the construction team.

Defects and snags are highlighted as site observations during the contract, and a final list is produced prior to practical completion. Defect lists would be signed off progressively up to practical completion.

But the work of the company does not stop there. A **defect liability period** occurs after the handover of a project during which a contractor must make good any outstanding defects.

## The project: St Bartholomew's Hospital

As part of the design brief, it was specified that all rooms in the hospital had to be maintained at specific temperatures, depending on the room type, at all times of the year.

Being an inner-city hospital, natural ventilation, through the use of windows that could be opened, and free or passive comfort cooling was not an option due to noise and pollution levels. The company had to design a completely mechanical air-conditioned and ventilated building.

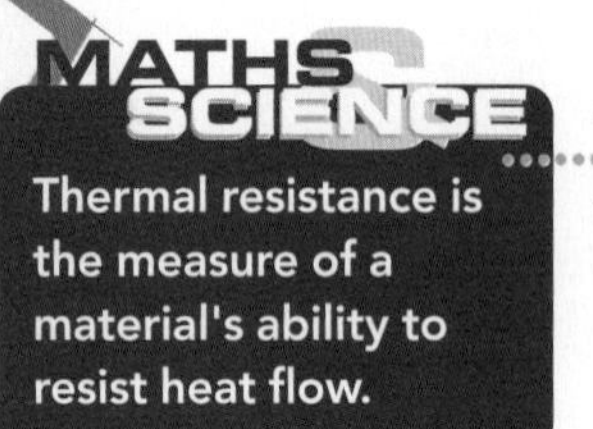

Thermal resistance is the measure of a material's ability to resist heat flow.

One of the first jobs for a building services engineer is to calculate the heat loss in cold weather and heat gain in hot weather for the building. This needs to take into account the differences between the desired indoor air temperature and the exterior temperature. For this they need to consider the daily temperature range, where the sun shines on the building, and the thermal resistance of walls, windows and roofs.

Together with all this other information they need to consider the uses and activities within the building. Heat gain can be reduced by the layout and orientation of the building. Building services engineers use energy and thermal modelling to advise clients and architects on the most energy-efficient facade and orientation of a building – depending on the scope of the project. The energy and thermal models are also used to determine the required size of the heating and cooling plant.

### External modelling

The first step in creating the solution is to create a model to investigate the thermal environment and the energy efficiency of the building. The starting point is the two-dimensional drawing provided by the architect. Using computers, TB+A is able to construct an accurate three-dimensional model showing the building structure and all windows and doors.

The three-dimensional model can be rotated to view the building from any angle.

The three-dimensional model is used to investigate the thermal environment and energy efficiency of the building. The engineer needs to consider what aspects affect the fabric and orientation of the building, and the look of the building **facade**.

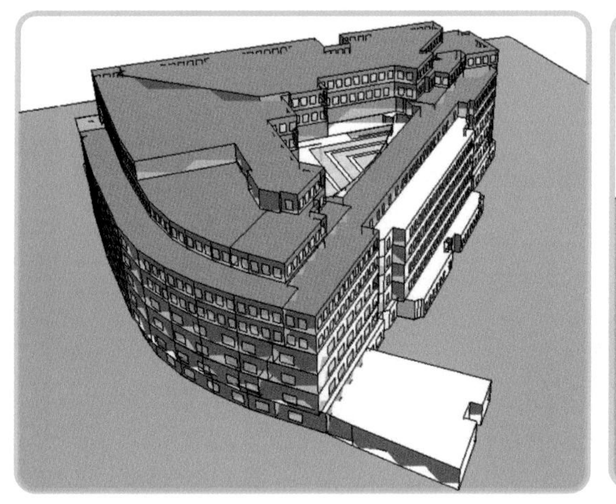

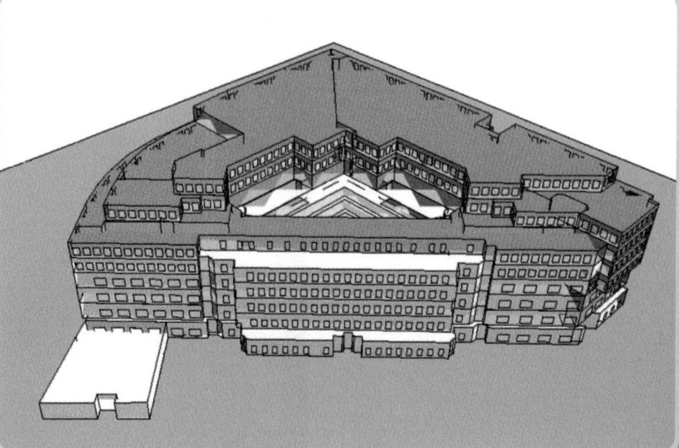

**facade** – the principal front of a building which faces on to a street or open space

As part of the process, adjacent buildings are reconstructed in the model. This allows the creation of a solar model. This model is run in real time and indicates how the sun moves around the building.

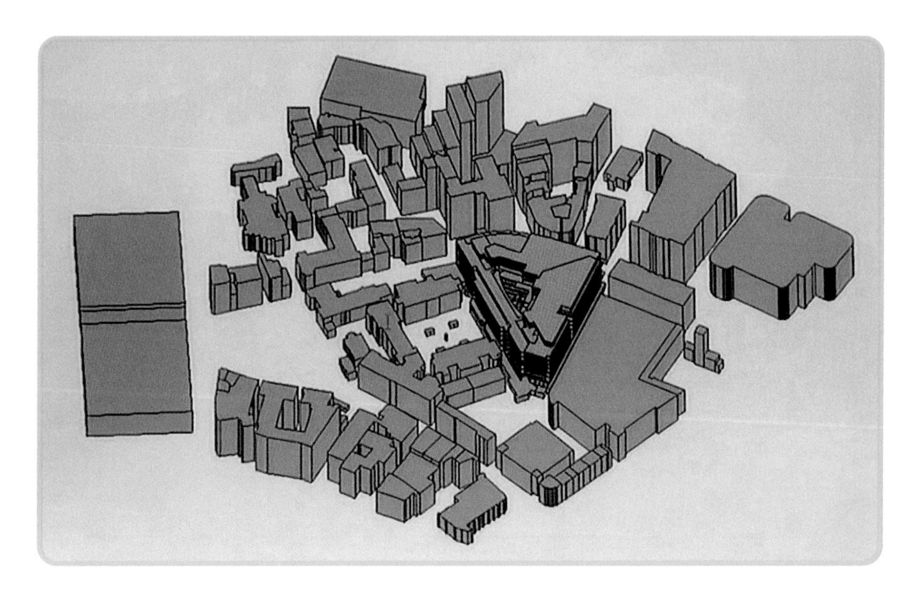

This will show how much shading is produced on the different faces of the hospital by the adjacent buildings.

The shape and surroundings of any building play a vital role in governing the energy consumption in that building. Such factors may cause heat gain when cooling is required and heat loss when gain is required.

It is important that modelling is used on the building as a whole. All the loads on each room (from the occupants and the equipment) are entered, the heating and cooling requirements are considered, and the solar aspect of the building, i.e. how the sun influences the interior temperatures, is included.

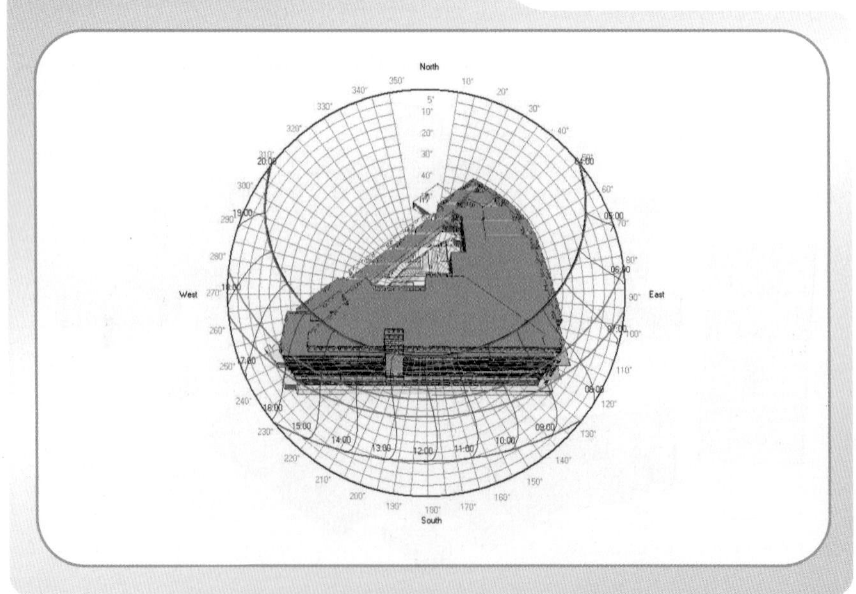


**FIG 8.5:** Map of sun's path around St Bartholomew's Hospital

Another important consideration is weather conditions. As part of the modelling process the average weather conditions for the last 20 years are input in to the model.

From the three-dimensional model the sun's path can be mapped as it travels around the building. A map is created for the whole year, allowing the engineer to consider different seasons and different times of the day.

Analysing the data showed that there was a large amount of solar gain on the higher floors of the hospital. To reduce the amount of solar gain, and thus reduce the amount of cooling required, the architect was advised to use external solar shading for the higher floors. The material used was **brise soleil**.

Model showing the addition of brise soleil and canopy structure

Brise soleil is a French term for a permanent sun shade used outside an external wall to provide solar shading to reduce glare and heat, including solar gain. Brise soleil are normally made up of horizontal or vertical brise soleil **louvres**, or may take the form of a decorative solar shading screen wall.

**louvre** – set of angled slats fixed or hung at regular intervals in a door, shutter or screen

Brise soleil may be needed to reduce solar heat gain through solar shading. It may also be required to prevent glare from the sun and localised thermal discomfort due to the sun shining on occupants, and to reduce or eliminate sky glare. Some shading devices may not combine all these roles effectively. For example, low-transmission glazing will reduce solar heat gain and sky glare, but has little effect on direct glare from the sun because of the high intensity of the sun's rays. Conversely, internal solar shading devices such as blinds can eliminate glare from the sky and sun but are less effective than brise soleil at reducing solar heat gain.

To ensure a different aesthetic, a canopy structure was used on one of the aspects of the hospital building, instead of the brise soleil. The canopy provides a shading effect in the same way.

The lower floors of the building were shaded by the adjacent buildings, so solar gain was not a factor here.

## Internal modelling

It is also important to consider the internal environment of a building.

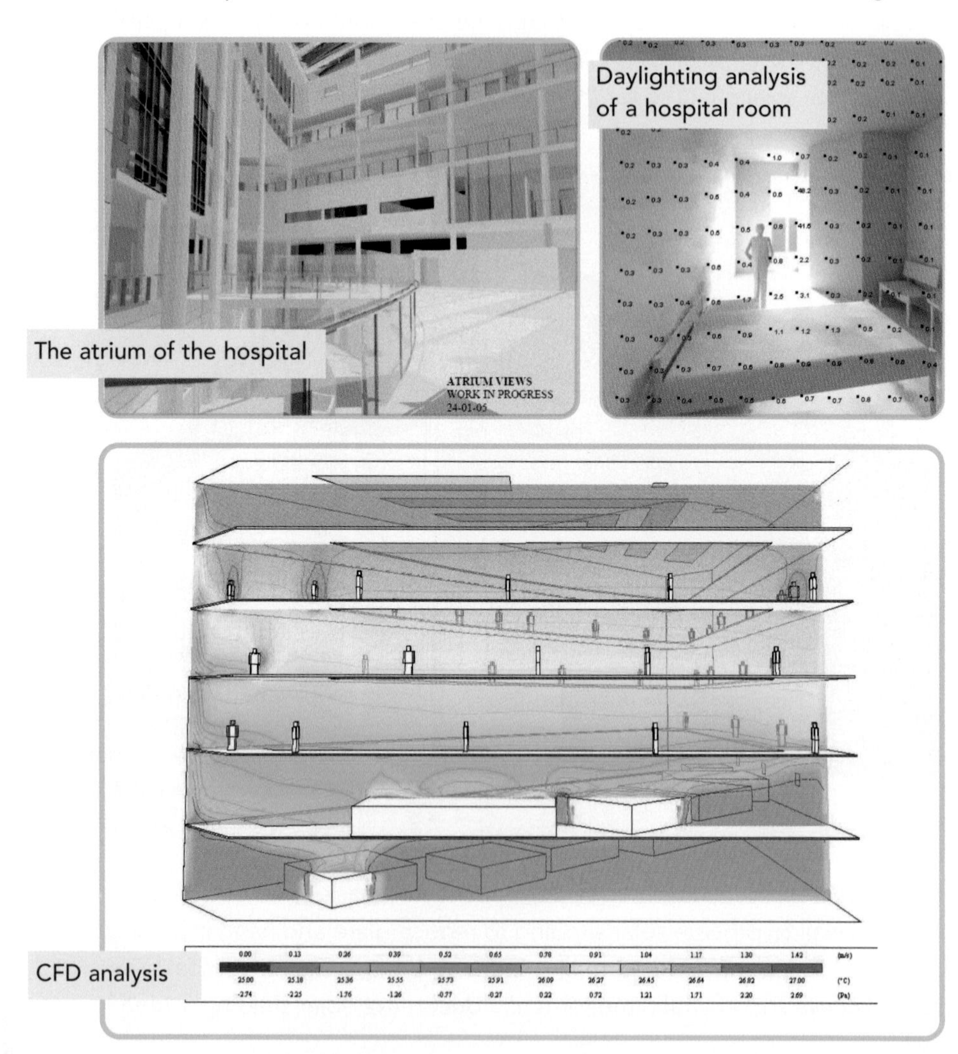


The atrium of the hospital

Daylighting analysis of a hospital room

CFD analysis

In the film you will see computational fluid dynamic (CFD) analysis of the central atrium of the building.

The design idea for the atrium, which was the main reception area to the hospital, was to use mix-modes cooling, using natural ventilation and the buoyancy of the air as it heats up, to be released from the top of the building.

It was important to ensure that, using this method, thermal comfort for the occupants in the atrium was achieved.

CFD analysis allows visualisation of the thermal map of the space, and allows the engineer to check that the required heating conditions are satisfied.

Another aspect of the design considered by TB+A was reduction of artificial lighting, thus reducing energy usage. Further details on lighting design and lighting efficiency are given in the lighting design case study. To see if a reduction in artificial lighting is a viable option, it is necessary to consider **daylighting** levels. Daylighting analysis maps daylighting intensity levels.

Within the overall architectural design of a building, particular attention is given to daylighting when the aim is to maximise visual comfort or productivity, or to reduce energy use. Energy savings from daylighting are achieved in two ways – either from the reduced use of electric lighting, or from passive solar heating or cooling.

## Design solution

Once the thermal conditions are understood, the engineer can begin to consider the requirements for the electrical, heating, cooling and ventilation systems necessary to maintain the specified thermal conditions.

It is necessary to produce plans and layout drawings of the services and the distribution routes. This will ensure that all the components of the system are able to fit.

# The career path

## PERSONAL PROFILE: Jonathan Foster

**Qualifications**
MSc Computational Fluid Dynamics
BEng Environmental Energy Engineering

**Professional affiliation**
Registered Building Research Establishment BREEAM Assessor

**Professional experience**
*Energy and Mechanical Consultant* — *Troup Bywaters + Anders*

Particular experience in building-physics analysis, such as:
- internal and external airflow analysis
- natural and mix-mode ventilation performance
- building energy consumption calculations/modelling
- summertime temperature predictions
- HVAC systems design and control.

**Projects**
- Nomura, Centurion House
- American Express, Belgrave House, Victoria
- Deloitte Building B, London
- Farringdon Row, Sunderland BREEAM

### Journey to environmental engineering

My interest in engineering was a gradual development. At school I was interested in the sciences and maths, but also drama and graphics.

I was originally planning on studying chemistry at university, but then decided that was not the subject for me. I thought that engineering would suit my abilities but I was unsure as to what mechanical engineers actually did.

At college I had developed an interest in the environment and energy efficiency. Research led me to a course on environmental energy engineering. After visiting the university, and talking to other students and the lecturers, I felt this course offered an interesting approach to what I thought engineering was.

**To find out more about a career in building services go to www.prospects.ac.uk and www.cibse.org.uk and www.tbanda.com**

APPENDICES

# Health & safety

## Occupational health and safety

### Facts and figures

In the UK there are 1.6 million workplace injuries every year, as well as 2.2 million cases of ill health caused or made worse by work.

Over 25 million working days are lost annually as a result of workplace accidents, injuries and ill health. Health and safety failures currently cost Britain's employers up to £6.5 billion every year.

### What is health and safety?

Health and safety is about preventing people from being harmed by work or from becoming ill by taking the right precautions, and providing a satisfactory working environment.

Because health and safety at work is so important there are rules which require all of us not to put ourselves or others at risk. The law is also there to protect the public from workplace dangers.

### Health and safety law

The basis of British health and safety law is the Health and Safety at Work etc. Act 1974.

This sets out the general duties which employers have towards employees and members of the public, and which employees have to themselves and to each other.

Under the legislation, employers must:

- make the workplace safe, without risks to health
- ensure plant and machinery are safe, and that safe systems of work are set and followed
- ensure articles and substances are moved, stored and used safely
- ensure that the workplace satisfies ventilation, temperature and lighting, and sanitary, washing and rest facility requirements
- provide adequate first-aid facilities
- give information, instruction, training and supervision
- ensure that appropriate safety signs are provided and maintained

- ensure that there is a correct and legal system for fully reporting accidents.

But it is not just the employer who has a responsibility for health and safety. The Health and Safety at Work Act places equal responsibility on the employee.

The employee must:

- cooperate fully in the event of any accident or fire
- report accidents and health and safety concerns to an appropriate person
- obey the accident procedures
- correctly use work items provided by the employer
- not interfere with or misuse anything provided for health, safety and welfare.

## Health and safety policy statement

As part of the legislation, it is necessary to display the company's health and safety policy statement.

**Helipebs (Holdings) PLC**

**Occupational Health and Safety Policy**

The Helipebs (Holdings) PLC's policy is to provide adequate control of the health and safety risks arising from our work activities. The Group is committed to the continual improvement of Occupational Health and Safety performance and Management System.

The Group will:

- Comply with all applicable OH&S legislation and other requirements of the organisation.
- Maintain safe and healthy working conditions.
- Provide and maintain safe plant and equipment.
- Ensure safe handling and use of substances.

Employees will, as appropriate, be made aware of their individual obligations relating Occupational Health and Safety.
Employees will be consulted on matters affecting their health and safety.

The Group will provide information, instruction and supervision for employees to ensure all employees are competent to do their tasks.
Employees will be trained in order to help prevent accidents, incidents and cases of work-related ill health, and improve Occupational Health and Safety performance.

Health and Safety Director – Helipebs (Holdings) PLC
G F Davis
G F Davis.

Health and Safety Officer – Helipebs Ltd
I Yemm

Health and Safety Officer – Helipebs Controls Ltd
R Williams
R J. Williams

Helipebs (Holdings) PLC
Sisson Road
Gloucester
GL2 0RE

Version: HHPLC OH&S Policy R 2.2 08-04-05

The health and safety policy document for Helipebs Controls Ltd is shown on page 185.

An employer with five or more employees should also have a written health and safety policy.

# Managing health and safety

## Risk assessment

A risk assessment identifies aspects of the working environment. It records what might pose a real and immediate danger to staff. A risk assessment on its own does not bring about health and safety protection. However, what it does provide is a list of places where risks exist and the action that needs to be taken to eliminate or reduce those risks.

In identifying and assessing risks in the workplace there are certain key questions that need to be addressed.

- What are the potential hazards?
- Who might be harmed by each identified hazard? How might they be harmed?
- What precautions already exist to prevent the risk?
- Can anything be done to reduce the amount of risk?

It is part of the law that the answers to these questions are recorded (if the employer has five or more employees). It is then necessary to implement the identified changes.

Every year or so, the risk assessment should be reviewed and updated if necessary.

## Common causes of accidents

About 60 per cent of fatal injuries to workers occur in construction, transport and storage, and in agriculture, forestry and fishing. The most common kinds of accident involved in fatal injuries are falling from a height, being struck by a moving vehicle, and being struck by moving or falling objects.

The highest cause of major incidents is slips and trips, often resulting in broken bones.

Areas of concern include:

- **Supervision and training**
  A lack of adequate supervision or training can lead to accidents, if workers disregard safety procedures or indeed are unaware of the safety procedures to begin with.

- **Working environment**

  Poorly guarded and badly maintained equipment and machines are obvious causes of injury. Accidents also occur by falling on, for example, slippery floors, poorly maintained staircases, or scaffolding.

  Noise, poor lighting and inadequate ventilation can also lead to accidents.

  An important factor that is often overlooked is that of the layout of the immediate work environment. A dirty, messy workspace can lead to problems and can increase the risk of fire, cross-contamination of products, etc. In the film you will notice that Ceramic Seals has a very clean working environment. This is due to the high levels of accuracy needed in its work.

- **Personal health and hygiene**

  Dirty surroundings and inadequate toilet and washing facilities can also lead to a lowering of personal hygiene standards, and cause health and safety issues.

  In the film you will notice companies such as Ginsters have very stringent washing and cleaning procedures. In cases such as this, not following safety procedures can lead to the contamination of food products.

- **Personal habits**

  On a basic level, elements such as tiredness cause many accidents in the workplace. Consider the possible consequences of an overtired person operating heavy machinery.

  Factors such as smoking in prohibited areas, where flammable substances may be present, also increase the risk of potential accidents.

## Accident prevention

The fact is that 70 per cent of workplace accidents could be prevented if employers put proper safety control measures in place.

- **Information, instruction and supervision**

  The health and safety policy must be displayed. Employees must also be informed of where they can go for health and safety advice.

  Consider the companies in the film. These companies display a large number of warning notices and instructional posters to promote quality and health and safety standards. The notices are displayed in very prominent positions to ensure continued reinforcement of the health and safety message.

  All employees must be given health and safety induction training when they start work, which should cover the basics such as first aid and fire safety. There should also be specifc health and safety training.

It is very important to ensure that employees are well trained in health and safety, but also have a positive attitude towards its implementation. This is not only to protect themselves but also to protect their workmates.

- **Safe plant and equipment**

  It is important that all plant and equipment that requires maintenance is identified, and that the maintenance is done.

  Machines and equipment must be regularly serviced and maintained by trained service engineers. Engineers must also ensure tools are correctly stored and are functioning efficiently.

- **Safe handling and use of substances**

  The risks from all substances hazardous to health must be identified and assessed. Hazardous substances not only include chemicals that people make or work with directly, but also dust, fumes and bacteria which can be present in the workplace.

  An important piece of legislation here is the Control of Substances Hazardous to Health Regulations 2002. In short, they are known as COSHH.

- **Personal protection**

  It is vital to ensure that people are given the correct personal protective equipment (PPE) whenever there are risks to health and safety that cannot be adequately controlled in other ways.

  The requirements for the provision and use of PPE at work are covered by the Personal Protective Equipment at Work Regulations 1992.

  PPE must be:

  - properly assessed before use to ensure it is suitable
  - maintained and stored properly
  - provided with instructions on how to use it safely
  - used correctly.

  In the film you will see a wide range of PPE being used. Items include face masks, ear protection, safety goggles and protective clothing, such as the protective coats used at Ginsters.

- **Working environment: workspace**

  A sign of a good engineer is a clean and tidy working area. Only the minimum number of tools for the job should be laid out at any one time. Tools and equipment should be laid out in a tidy and logical manner. Any tools not immediately being used should be cleaned and stored away.

## Health and safety and the law

Apart from the Health and Safety at Work Act there are a large number of other laws that protect engineers.

These include:

- Management of Health and Safety at Work Regulations 1999 – require employers to carry out risk assessments, make arrangements to implement necessary measures, appoint competent people and arrange for appropriate information and training
- Personal Protective Equipment at Work Regulations 1992
- Provision and Use of Work Equipment Regulations 1998 – require that equipment provided for use at work, including machinery, is safe
- Manual Handling Operations Regulations 1992 – cover the moving of objects by hand or bodily force
- Control of Noise at Work Regulations 2005 – require employers to take action to protect employees from hearing damage
- Electricity at Work Regulations 1989 – require people in control of electrical systems to ensure they are safe to use and are maintained in a safe condition
- Control of Substances Hazardous to Health Regulations 2002.

There are also specific regulations that cover particular areas:

- Construction (Design and Management) Regulations 2007 – cover safe systems of work on construction sites
- Gas Safety (Installation and Use) Regulations 1998 – cover safe installation, maintenance and use of gas systems and appliances in domestic and commercial premises.

There are also laws to protect the environment such as:

- Rivers (Prevention of Pollution) Act 1961
- Control of Pollution (Amendment) Act 1989
- Clean Air Act 1993
- Environmental Protection Act 1990.

An engineer is not expected to have detailed knowledge of all this legislation, but they are expected to know that it exists. They are also expected to know the main topic areas that these regulations cover, and how they affect their own working conditions and their responsibilities.

# Standards: ISO

## Why do we need standards?

If there were no standards, would we notice? Standards affect most aspects of our lives – although this effect is only noticed when there is an absence of standards, or the standards themselves are deficient. For example, when purchasing products we soon notice if those products are of poor quality, or are not fit for purpose. When products meet our expectations, we tend to take this for granted.

Standards ensure that products and services conform to specific requirements, e.g. quality and safety.

## What is ISO?

ISO (International Organization for Standardization) is an international body that sets and develops technical standards for various industries.

The term ISO is derived from the Greek 'isos', meaning 'equal'.

First established in 1947 as a non-governmental organisation, ISO represents a network of national standards institutes from over 150 different countries, working in partnership with organisations, governments and industries around the world.

ISO standards are developed by technical committees comprising over 30,000 qualified representatives from industry, research institutes, governments and consumer bodies from across the world, who come together as equal partners in the quest for global standards.

Between 1947 and 2007, ISO published more than 16,000 standards. Its work programme ranges from standards for agriculture and construction through to the newest information technology developments. Standardisation of engineering materials, such as the size of screw threads, is a result of the work of the ISO.

Many of the case study companies have achieved ISO standards. One of those standards, achieved by Helipebs Controls Ltd, is ISO 9001.

### ISO 9000: management standards

ISO 9000 is a group of standards that have been developed to provide a framework around which a quality management system (QMS) can effectively be implemented.

ISO 9001 is the requirements standard, and is intended for use in any organisation which designs, develops, manufactures, installs and/or services any product, or provides any form of service. The standard details a number of requirements that an organisation needs to fulfil in order to gain accreditation.

# Certificate of Registration

**QUALITY MANAGEMENT SYSTEM - ISO 9001:2000**

*This is to certify that:*

**Helipebs (Holdings) PLC**
**Sisson Road**
**Gloucester**
**GL2 0RE**
**United Kingdom**

*Holds Certificate No:* **FM 14317**

*and operates a Quality Management System which complies with the requirements of BS EN ISO 9001:2000 for the following scope:*

The design and manufacture of a range of hydraulic and pneumatic cylinders, valves, servo-control equipment and accessories. These products may also be manufactured under licence or to customer specifications.
The manufacture of electronic sub-assemblies to customer specification.

*For and on behalf of BSI:*

*Managing Director, BSI Management Systems (UK)*

Originally registered: **17/01/1992** Latest Issue: **10/07/2006**



*Page:* 1 of 2

BSi
Management Systems

This certificate was issued electronically and remains the property of BSI and is bound by the conditions of contract.
This certificate does not expire. An electronic certificate can be authenticated online.
Printed copies can be validated at www.bsi-global.com/ClientDirectory

The British Standards Institution is incorporated by Royal Charter.
Management Systems (UK) Headquarters: P.O. Box 9000, Milton Keynes MK14 6WT. Tel: 0845 080 9000

The importance of the ISO 9000 series of standards is that it provides assurance of quality management. This means what the organisation does to enhance customer satisfaction by meeting both customer and applicable regulatory requirements. When someone purchases a product from an organisation which has ISO 9001 certification, the customer has the assurance that the quality of product he or she receives is as expected.

In fact, ISO 9000 is emerging as a global standard by which organisations are judged for the quality of their goods and services. It can help manufacturing companies to achieve standards of quality that are recognised and respected throughout the world.

More than half a million organisations in more than 140 countries are implementing ISO 9000.

## ISO 14000: environmental standards

ISO 14000 is a series of standards that deal with environmental impact management. The ISO 14000 standard provides guidance for the development and implementation of an environmental management system (EMS). This means what the organisation does to minimise the harmful effects on the environment which may be caused by its activities.

# Certificate of Registration

**ENVIRONMENTAL MANAGEMENT SYSTEM**

*This is to certify that:*

**Helipebs (Holdings) PLC**
**Sisson Road**
**Gloucester**
**GL2 0RE**
**United Kingdom**

*Holds Certificate No:* **EMS 69317**

*and operates an Environmental Management System which complies with the requirements of BS EN ISO 14001:2004 for the following scope:*

The manufacture of hydraulic and pneumatic cylinders, valves, and electrical control systems. The manufacture of grinding media.

*For and on behalf of BSI:*

*Managing Director, BSI Management Systems (UK)*

Originally registered: **05/03/2004** Latest Issue: **13/01/2006**

*Page:* 1 of 2

This certificate was issued electronically and remains the property of BSI and is bound by the conditions of contract.
This certificate does not expire. An electronic certificate can be authenticated online.
Printed copies can be validated at www.bsi-global.com/ClientDirectory

The British Standards Institution is incorporated by Royal Charter.
Management Systems (UK) Headquarters: P.O. Box 9000, Milton Keynes MK14 6WT. Tel: 0845 080 9000

It applies to those environmental aspects which the organisation can control and over which it can be expected to have an influence.

This standard is applicable to any organisation that wishes to:

- implement, maintain and improve an environmental management system
- assure itself of its conformance with its stated environmental policy
- demonstrate such conformance to others
- seek certification/registration of its environmental management system by an external organisation.

A number of the case study companies have an effective EMS, which helps companies manage the impact of their business activities on the environment.

Examples of EMS implementations include recycling programmes, hazardous materials management, and reduction in emissions.

For many companies, ISO 14001 certification is a requirement before they are able to get work contracts. This is common in both the United States and the EU. Companies that meet the ISO 14001 standard have a competitive edge in the market by showing concern for environmental issues.

With the recent focus on how climate change affects polar ecosystems, the ISO 14000 series of standards remains highly relevant. In fact, ISO has a new standard (ISO 14064) for the quantification, reporting and verification of greenhouse gas emissions.

## ISO/TS 16949: automotive standards

ISO/TS 16949 is a technical specification which specifies the quality system requirements for the automotive industry supply chain. Along with ISO 9001, ISO/TS 16949 provides guidelines for the design, development, production, installation and servicing of automotive-related products.

ISO/TS 16949 was developed by the International Automotive Task Force (IATF) and has become a mandatory set of requirements for automotive companies in both North America and Europe.

# Index


ABS (antilock brakes) 33
absolute zero 143
absorption 99
acceleration 38
accent lighting 105
acceptance of order 19
accidents, common causes of 186–7
active safety features 33
actuators (robotics) 119
additive colours 56, 98
air conditioning 171–3, 176
aircraft, fluid power used in 8
alloys 146, 148
aluminium oxide/alumina 145
amorphous materials 144
ampere (A) 104
angles of projection 79–81
appliances, domestic 165
apprenticeships 157
artwork 67–8
assembly drawings 83
atoms 139–42
  in insulators 168
  and thermal expansion 143
autochangers 131
automated production 129–31
  *see also* robots
automotive industry *see* motor industry
auxiliary views 82–3

bakelite 171
ball bearings 15, 128, 129
bar (pressure unit) 15
bearings 15, 128–9
bevel gears 126
binding (printing) 62, 69–70
bleed (printing) 61
blueprints 94
body-centred cubic (BCC) 140
bonding 139, 141, 143
bonnets, car 40–2, 43, 44
brakes 12, 13, 15
  ABS (antilock brakes) 33
brazing 146, 148
  testing 153
break-even point 89
brief 84
  building services 174
  fluid power projects 18
  printing jobs 66–7
brise soleil 179
brittleness 36, 37
bubblejet printers 65
buckle folders 59–60
building envelope 166–7
building services 94
  lighting design 94–6, 105–13, 181
  mechanical design 160–8, 171–82
building services engineers 160
buildings
  electricity in 103–5
  energy use in 164–5
  fabric of 167–8
business failure 75, 85–6

cables in buildings 103–4
CAD (computer-aided design) 85, 161
cams 17
candela (cd) 101–2
capital expenditure 67
carbon dioxide emissions 162, 163–4
  and buildings 95–6, 162, 165
cars 8, 15
  brakes 12, 13, 15, 33
  safety 30–1, 33, 38–46
  structure of 33–6
  *see also* motor industry
ceramics 138–9, 143–6
certificate of conformity 24
CFCs (chlorofluorocarbons) 163, 173
charge
  on atomic particles 141
  and electrostatics 64
chartered engineers 74
chassis, vehicle 34, 35
chemical symbols 145
chlorophyll 99
civil engineers 91
climate change 95, 163–4
CMYK four-colour process 55
CNC (computerised numerical control) machines 150–1
cobalt 148
coefficient of expansion 143–4
  in applications 147, 148
coefficient of friction 15, 16
cold welds 15
collisions *see* crashes
colour 53–6, 98–9
  in retail outlets 110–11, 112
colour rendering 112
colour temperature 110–11
comb binding 62
commercial viability 85
composites 146
compound gear trains 125
compound gears 124–5
compounds 140
computational fluid dynamic (CFD) analysis 180–1
computer-aided design (CAD) 85, 161
computer modelling
  in building projects 176–81
  and crash test dummies 44
  in lighting design 108
computer simulation 44
computer-to-plate 57

computers
  in building services 161, 180–1
  CAD 85, 161
  CNC machines 150–1
  and colours 56, 98
  and digital devices 123
  image resolution 54
  in printing 53, 57, 63–5, 67–8
conduction 166, 167
conductivity 139, 142, 143
conductors 103–4
conduit 104
conservation of energy 39
consultancy 95, 161
contone (continuous tone) 53
contour charts in lighting 108
contraction 143
contracts, winning 174, 175
convection 166
conveyors 126–8
  in food production 130–1
cooking, automated 129–31
cooling spirals 130–1
coordinate measuring machine 155–6
corrosion resistance 36, 37, 139
costs 88–90
  of ceramic raw materials 148
  and CNC machines 151
  and planning print jobs 67
  and quality 25–6
  of robots 118
  and waste 132
covalent bonds 139, 141
  and thermal conductivity 143
crash test dummies 44
crashes, car 38–42, 44–6
creep (printing) 61
crumple zones 39
cryogenic conditions 137
crystals 138–9, 146
  polymer 168
current 104
  and spot welding 37
customer needs 20, 85
cutaways 13
cutting plane 81–2

daylighting analysis 180, 181
defect liability period 176
deflection and structure 35–6
deforestation 162
degrees of freedom (robotics) 119
demand 85
design
  CAD (computer-aided) 85, 161
  cars 30, 31, 43–4, 85–7
  ceramic products 149–50
  fluid power projects 18–20
  lighting design 94–6, 105–13, 181
  mechanical design 160–8, 171–82
design engineers 27
  lighting 108–13
design envelope 19
design parameters 19
diamond, lattice of 141
diffuse lights, use of 111
diffuse reflection 100
diggers, fluid power and 8
digital devices 123
digital printing 63–5
distribution channels 117
downlighting 105
drawings 76–83
  ceramic products 149
  hydraulic systems 19–20
ductility 36, 37, 142

eccentric rotating cylinders 17
economy, the 50, 74
effort 120, 121, 122
elastic deformation 36
electrical conductivity 139, 142
electrical energy 104
electrical engineers 160
electricity 103–5
  and fossil fuels 165
  and lighting systems 95
electromagnetic radiation/waves 97–8
  and heat transfer 166
  infrared 99, 103, 163
  *see also* light
electromotive force units 104
electrons 139, 140, 141
  and heat conduction 166
  in insulators 168
  in metal lattices 142
electrostatic printing 64–5
electrostatics 64
elements 140, 141
elevation drawings 81
elevators, fluid power and 8
emission 56, 99
energy
  and absorption of light 99
  and buildings 162, 164–5, 167, 181
  in car crashes 38–9
  conservation of 39
  and cooking 129
  electrical 104
  fluid power transmits 8
  and greenhouse effect 164
  and heat conduction 166
  and lighting systems 95–6, 110
  printing consumes 52
  and thermal expansion 143
  and waves 97
  and work 129
engineering drawings *see* drawings
engineering materials 138
  *see also* ceramics; metals
engineers 47, 74
  building services 160
  civil engineers 91
  design engineers 27
  environmental energy 182
  head of engineering 133
engines, car 34
environment
  climate change 95, 163–4
  ISO standards 10, 52, 189, 192–3

epoxy resin 171
equilibrium 120
  thermal 166
equipment, safety 188
equity partnerships 161
expansion 143–4
  applications 147, 148
exploded views 83

face-centred cubic (FCC) 140
facing (metal removal) 21, 22
farm equipment, fluid power and 8
feasibility studies 174–5
feature lighting 106
feedback 119
ferromagnetic materials 148
films, printing with 56–7
finishing (printing) 58
firing process 146
first angle of projection 79–81
fixed costs 88, 89, 90
flexography printing 58
fluid power 8–27
fluids 11
  for hydraulic systems 23
  *see also* gases; liquids
folding paper 59–60, 61–2
food manufacturing 116–18, 126–33
forces 120
  and car crashes 38, 39
  and friction 14–15
  and hydraulic cylinders 12–13
  and laws of motion 38
  and levers 120–2
  structures transmit 33
  turning forces 120
forecasts 84
Formica 171
fossil fuels 162, 164, 165
four-colour process 55
frequencies 97, 98
friction 14–15, 16
  bearings reduce 128
frontal offset crash tests 45
FTP (file-transfer protocol) 68, 161
fulcrum 120, 121, 122

gamma rays 97
gases
  convection in 166
  greenhouse 96, 163, 164
  in pneumatics 8
GDP (gross domestic product) 74
gear pitch 125
gear pumps 17
gear trains 124
  compound 125
gearing ratios 123–4
gears 123–6
general arrangement drawings 19
glare 105, 179
glass fibres in composites 146
global warming 163–4
GNP (gross national product) 94
grain, paper 61–2
graphs, linear 88, 89
gravure printing 58
green plants and light 99
greenhouse effect 163–4
greenhouse gases 96, 163, 164
gross value added 116
grounded conduit 104

halftone 53–4
halocarbons 163
hardness 36
hatching (drawings) 82
HCFCs (hydrochlorofluorocarbons) 163, 173
head injuries 40, 41–2
health and safety 184–9
heat
  and absorption of light 99
  and greenhouse effect 164
  light bulbs give out 103
  and thermal expansion 143
heat gain (buildings) 166–7, 168, 176, 178, 179
heat transfer 166
heating systems 171–3
helical gears 126
helium 152
hexagonal close packing (HCP) 140
HFCs (hydrofluorocarbons) 163
high-care areas 132
horizon in perspective drawings 78
hot spots (lighting) 109
HVAC 171–3
hydraulic cylinders 8, 12–15, 19
hydraulic multiplication 12–13
hydraulic pressure, applying 15–17
hydraulic pumps 15, 17
hydraulics 8
  *see also* fluid power
hygiene, food 132

illuminance 102
images 53–4, 56–8
imperial units 10, 54
imposition (printing) 68
in-line ovens 129–30
incandescent lamps 165
  *see also* light bulbs
incidence, angle of 99–100
incorporated engineers 74
index of refraction 101
industrial ceramics 138–9
inert gases in light bulbs 103
inertia 38
infrared 99, 103, 163
inkjet printers 65
input 118–19
installation drawings 19
insulation 165, 168
insulators 103–4, 148, 168
internet
  building services use 161
  in printing 68
ionic bonds 139, 141–2
ions 139
  in ionic lattices 141–2
iron 148
ISDN 161
ISO 9000 series 10, 138, 190–2

ISO 14000 series 10, 52, 192–3
ISO standards 190–3
automotive 183
environment 10, 52, 192–3
paper sizes 62
quality 10, 138, 190–2
isometric drawings 78–9, 81
IT (information technology) 161
*see also* computers

job bag 67
just in time (JIT) production 148–9

kinetic energy 143, 166
kinetic friction 15
knife folders 60
Kyoto Protocol 96

laser printers 64
lathes 20–2
lattices 140, 141–2
laws *see* legislation/ regulations
leak detection 152
LEDs (light-emitting diodes) 110
legislation/regulations
building performance 162
car safety issues 43, 45
construction 175
health and safety 184–5, 188, 189
leverage 120
levers 120–2
lifts (elevators), fluid power and 8
light 97–102
absorption 99
and colour 55, 56, 98–9
exposing negatives with 57
reflection 55, 99–100
refraction 98, 100–1
speed of 98, 100–1
light bulbs 102–3
*see also* incandescent lamps
light fittings 106–7
lighting design 94–6, 105–13
to save energy 95–6, 181
lighting design engineers 94, 108–13
linear graphs 88, 89
liquids
convection in 166
in hydraulics 8
load 120, 121, 122
low-care areas 132
lumen (lm) 102
luminaires 106–7
luminance 101
luminous flux 102
lux (lx) 102

magazines, printing 63, 66–70
magnetic materials 148
maintenance
lighting systems 112
and safety 188
malleability 36, 37, 142
Management Information Systems (MIS) 67
margins, profit 110
market research 84
mass
and acceleration 38
and momentum 39
mass number 141
mass spectrometers 152–3
materials
for building 168–9
ceramics 138–9, 143–6
engineering 138
and friction 15, 16
for hydraulic systems 23
metals *see* metals
properties 36–7, 142–5
mechanical advantage 122
mechanical design 160–8, 171–82
mechanical engineers 160
mechanical properties 36–7, 142
medium, wave 97
melamine 171
melting point 143
metals 136–7, 138
brazing to ceramics 146
in compounds 139
as conductors 103, 148
expansion 143, 144, 147, 148
lattices 142
properties 36–7, 147–8
methane 132, 163
metric units 10, 54
minerals, engineering with 136
mock-ups, lighting design 108
modelling *see* computer modelling
molecules and thermal expansion 143
momentum 39
motion
and crashes 38–42, 44–6
gears transmit 123
Newton's laws of 38
in robotics 123
motor industry
and the economy 74
standards in 183
*see also* cars

nanotechnology 136
natural gas 132
negatives, film 56–7
neutrons 140, 141
newspapers, printing 63
Newton, Sir Isaac 56, 98
Newton's laws of motion 38
newtons (N) 12
niche markets 76
nickel 147–8
non-metals in compounds 139
nucleus, atomic 140, 141
nuggets (spot welding) 37–8
nylon 170

oblique drawings 78, 79
offset lithography 56–8
ohm 104
on-demand printing 63
one-off service 76
one-point perspective 78

open-loop control 123
order confirmation 67
orthographic projection 79–81
  auxiliary views in 82–3
output 118–19
ovens, commercial 129–30
overheads, print industry 67
ozone 163

packing, automated 131–2
page layout software 68
Pantone colour matching system 56
paper folding 59–60, 61–2
paper sizes 62–3
partnerships, equity 161
parts lists/tables 77, 83
Pascal, Blair 11
pascal (Pa) 15
Pascal's Law 11, 12
passenger cells 34–5
passive bumper system 41
passive safety features 33
patents 43
pedestrians 30–1, 33, 40–4
penalty clauses 26
perfect binding 62
perspective drawings 77–8
Perspex 170
petroleum 169
photochemical reactions 57
photocopiers 64
physical properties 142, 143–4
pie charts 164–5
piezoelectric printers 65
piston pumps 17
pistons 12–13, 15
pitch, gear 125
pitch circle 125
pitch diameter 125
pivot 120
pixelation 54
pixels 54
plan drawings 81
planes, cutting 81–2
planning
  automated production lines 129
  building services 174–6
  ceramic products 148–50
  fluid power projects 18–20
  pedestrian safety issues 43–4
  print industry jobs 66–7
  small engineering companies 84–6
plant, safety issues in 188
plastic deformation 36, 139
plastics 168–71
  in composites 146
  engineering with 138
  thermal conductivity 143
  *see also* polymers
plates, printing 56–8, 68–9
plug and socket, ceramic 147–8
pneumatics 8
  *see also* fluid power
policy statements, health and safety 185–6
pollution 189
  and printing 52
  *see also* environment
polyester resin 171
polymers 168, 169
  *see also* plastics
polypropene 170
polystyrene 170
polythene 170
pop-up bonnet 41–2, 44
power 104
  and fluids 11
  *see also* fluid power
pre-flight artwork 68
pre-press 51, 68
precision engineering 136
press pass 69
presses 51–2
pressure 11
  applying, pumps for 15–17
  testing hydraulic components 24
pricing 67, 148
primary colours 55–6
print industry 50–71
print minders 69
printers 64–5
printing plates 56–8, 68–9
prisms, light through 98
process (robotics) 118–19
product development 84
  food and drinks industry 116
product portfolio 117
production
  ceramic products 150–1
  hydraulic systems 20–1
  motor industry 44
  print industry 67–70
  small engineering company 90
profit margins 110
proofs 68
protective equipment 188
protons 140, 141
prototypes
  motor industry 44
  small engineering company 90
psi (pressure unit) 15, 24
PTFE 170
public health engineers 160
pumps, hydraulic 15, 17
PVC 170

quality
  in construction 175
  in hydraulics manufacturing 25–6
  ISO standards 10, 138, 190–2

rack ovens 130
rack and pinion gears 126
radiation 166, 167
  *see also* electromagnetic radiation
rainbows 98, 101
ratios, units for 122
raw materials, ceramic 145–6, 148
rays 100
reception areas, design of 108–10
reciprocating motion 126
reflection 55, 99–100
refraction 98, 100–1
registration, print 69
rejects *see* scrap
relief printing 58

research, food industry 116
resistance, unit of 104
resistors in printers 65
resolution 54
retail stores, design of 110–12
RGB (Red Green Blue) 98
risk assessment 186
robots 118–19, 120, 121
  in food manufacturing 118, 129, 131
  moving arms 123–6
  reduce costs 90
  spot welding by 37–8
roll bars 35
roll hoops, MG 86–7
roller bearings 129
route cards 149–50

saddle stitching 62
safety
  cars 30–1, 33, 38–46
  health and safety 184–9
sales channels 90
scale 77
scattering 100
schematic drawings 14
scrap/rejects
  CNC machines reduce 151
  in food manufacturing 132
  in hydraulics manufacturing 25, 26
screen printing 59
sealing components 15
seat belt holder, MG 87
seat belts 38, 39
sections (drawings) 81–3
sections (printing) 70
sensors (robotics) 119
shading 178, 179–80
sheet-fed press 51, 63
shunting (spot welding) 37
side impact tests 44, 46
simulated tests 43
sintering 146
sketching 77–9
sodium chloride 141, 143
solar panels 99
space shuttle 144–5
spectra 55, 97, 98
speed 125
  and gears 125
  of light 98, 100–1
  *see also* velocity
spot colour 56
spot welding 37–8
spur gears 126
standards *see* ISO standards
stapling 62
step down transformers 105
stepper motors 123
steradian 101
stiffness, ceramic 142
stock 62
strength 36, 153–4
structural integrity 31, 87
substances, safe handling of 188
subtractive colours 55, 98–9
sun
  and building heating 178, 179
  and greenhouse effect 163, 164
suppliers 90
  in just in time production 149
supply, electricity 104–5
surface thickness testing 154–5
suspension, car 33

take-up screws 127, 128
task illuminance 108, 112
task lighting 105
teeth, gear 123, 125
temperature 111
  absolute zero 143
  and the greenhouse effect 163–4
templates 76–7
tendering 174, 175
tensile strength 36, 153–4
Terylene 170
testing
  ceramic components 152–6
  food industry 132
  hydraulic components 24
  motor industry 43, 44–6
  print industry 67–70
  small engineering company 90
thermal bubble printers 65
thermal conductivity 139, 142, 143
thermal equilibrium 166
thermal expansion 143–4
  applications 147, 148
thermal properties 143
thermal resistance 176
thermal storage 168
thermoplastics 169–70
thermosetting plastics 169, 170–1
thin-walled structures 35–6
third angle of projection 79–81
thread section sewing 62
three-dimensional objects 77
three-point perspective 78
thrust, car 33–4
tolerance 21, 23, 151–2
  measuring 156
toner 65
torque 33–4
toughness 36, 37
traceability, product 138, 150
traction control 33
traditional ceramics 138
transformers 105
transmission, car 33
tungsten 103
turning forces 120
turning techniques 21
two-point perspective 78

units 10
  of electrical energy 104
  for light measurements 101–2
uplighting 106

vane pumps 17
vanishing points 78
variable costs 88–9
variable data printing 63–4
velocity 38, 100
  and momentum 39
  *see also* speed
velocity ratio (levers) 122

ventilation 171–3, 176, 180
virtual build 44
visible light 55, 97, 98
  *see also* light
volt (V) 104
voltages, electricity supply 104–5

wall washing (lighting) 106
waste *see* scrap/rejects
watt (W) 104
wavelength 97, 98
waves 97, 98
wear resistance 139
web-enabled printing 64
web printing machines 63
web-to-print 64
welding, spot 37–8
white light 56, 98
windbreak screen, MG 87
windows 168
wire binding 62
wires and cables 103–4
wiring circuits 105
work 39, 104, 129
working drawings 79–81
working envelope (robotics) 119
worm gears 126

X-rays 154–5

**Acknowledgements**
The Author and the Publishers wish to thank the following for permission to reproduce photographs:

**Alamy** p126 – Bevel gear (Picturesbyrob); p126 – Helical gear (Charles Stirling); p126 – Rack and pinion gear (Rimmer); p126 – Spur gear (Andres Rubtsov); p126 – Worm gear (Tony Lilley); p128 – Ball bearing (David J. Green); p129 – Roller bearing. **NASA** p144 – Space shuttle; **Roger Scruton** p121 – Wheelbarrow. **Stanley Tools** p120 – Pliers. **Stephen Moulds** p53 – Laura half-tone; p54 – Balscote church.

All other images appear courtesy of the companies cited in the case studies and are reproduced from the *Engineering@work DVD Resource Pack*.